FABRIC DYEING AND MAKING TEXTILE COLORING MIXTURES

CLASSIC METHODS, MATERIALS, AND RECIPES
FOR OLD-TIME CLOTH COLORS

BY **PAUL N. HASLUCK**

ORIGINALLY PUBLISHED IN 1906

LEGACY EDITION

HASLUCK'S TRADITIONAL SKILLS LIBRARY
BOOK 7

Doublebit Press
Eugene, OR

New content, introduction, and annotations
Copyright © 2020 by Doublebit Press. All rights reserved.

Doublebit Press is an imprint of Eagle Nest Press
www.doublebitpress.com | Eugene, OR, USA

Original content under the public domain. Originally published in 1906 by Paul N. Hasluck under the title COLOURING MATTERS FOR DYEING TEXTILES.

This title, along with other Doublebit Press books including the Hasluck's Traditional Skills Library, are available at a volume discount for youth groups, outdoors clubs, or reading groups.

Doublebit Press Legacy Edition ISBNs
Hardcover: 978-1-64389-072-2
Paperback: 978-1-64389-073-9

Disclaimer: Because of its age and historic context, this text could contain content on present-day inappropriate methods, activities, outdated medical information, unsafe chemical and mechanical processes, or culturally and racially insensitive content. Doublebit Press, or its employees, authors, and other affiliates, assume no liability for any actions performed by readers or any damages that might be related to information contained in this book. This text has been published for historical study and for personal literary enrichment toward the goal of preserving the American handcraft tradition, timeless trade skills, and traditional artisanal knowledge.

First Doublebit Press Legacy Edition Printing, 2020

Printed in the United States of America
when purchased at retail in the USA

INTRODUCTION
To The Doublebit Press Legacy Edition

The old experts of artisanal trades, country and homestead knowledge, and the woods and mountains taught timeless principles and skills for centuries. Through their timeless books, the old experts offered rich descriptions of how the world works and encouraged learning through personal experiences *by doing*. Over the last 125 years, manufacturing, farming, and construction have substantially changed. Of course, many things have gotten simpler as equipment and technology have improved. In addition, some activities of pre-digital times are now no longer in vogue, or are even outright considered inappropriate or illegal. However, despite many of the positive changes in manufacturing and crafting methods that have occurred over the years, *there are many other skills and much knowledge that have been forgotten.*

By publishing *The Hasluck Traditional Skills Library*, it is our goal at Doublebit Press to do what we can to preserve and share the works from forgotten teachers that form the cornerstone of the history of the American artisans and traditional crafts. Through remastered reprint editions of timeless classics, perhaps we can regain some of this lost knowledge for future generations.

This book is an important contribution traditional handcraft and country skills literature and has important historical and collector value toward preserving the American handcraft and outdoors tradition. The knowledge it holds is an invaluable reference for practicing skills and hand craft methods. Its chapters thoroughly discuss some of the essential building blocks of knowledge that are fundamental but may have been forgotten as equipment gets fancier and technology gets smarter. In short, this book was chosen for Legacy Edition printing because much of the basic skills and knowledge it contains has been forgotten or put to the wayside in trade for more modern conveniences and methods.

With technology playing a major role in everyday life, sometimes we need to take a step back in time to find those basic building blocks used for gaining mastery – the things that we have luckily not completely lost and has been recorded in books over the last two centuries. These skills aren't forgotten, they've just been shelved. *It's time to unshelve them once again and reclaim the lost knowledge of self-sufficiency.*

Based on this commitment to preserving our outdoors and handcraft artisanal heritage, we have taken great pride in publishing this book as a complete original work. We hope it is worthy of both study and collection by outdoors folk in the modern era of outdoors and traditional skills life.

Unlike many other photocopy reproductions of classic books that are common on the market, this Legacy Edition does not simply place poor photography of old texts on our pages and use error-prone optical scanning or computer-generated text. We want our work to speak for itself, and reflect the quality demanded by our customers who spend their hard-earned money. With this in mind, each Legacy Edition book that has been chosen for publication is carefully remastered from original print books, *with the Doublebit Legacy Edition printed and laid out in the exact way that it was presented at its original publication.* We provide a beautiful, memorable experience that is as true to the original text as best as possible, but with the aid of modern technology to make as beautiful a reading experience as possible for books that can be over a century old.

Because of its age and because it is presented in its original form, the book may contain misspellings, inking errors from print plates, and other printing blemishes that were common for the age. However, these are exactly the things that we feel give the book its character, which we preserved in this Legacy Edition. During digitization, we ensured that each illustration in the text was clean and sharp with the least amount of loss from being copied and digitized as possible. Full-page plate illustrations are presented as they were found, often including the extra blank page that was often behind a plate. For the covers, we use the original cover design to give the book its original feel. We are sure you'll appreciate the fine touches and attention to detail that your Legacy Edition has to offer.

For traditional handcrafters and classic artisanal enthusiasts who demand the best from their equipment, this Doublebit Press Legacy Edition reprint was made with you in mind. Both important and minor details have equally both been accounted for by our publishing staff, down to the cover, font, layout, and images. It is the goal of Doublebit Legacy Edition series to be worthy of collection in any outdoorsperson's library and that can be passed to future generations.

Every book selected to be in this series offers unique views and instruction on important skills, advice, tips, tidbits, anecdotes, stories, and experiences that will enrich the repertoire of any person who enjoys escaping a bit from today's modern technology-based, cookie-cutter, and highly industrialized skills. Instead, folks seeking to make things with their hands like the old days may find great value from these resurrected instructional manuals from the past. These books were not simply written to be shelved in a library – they contain our history and forgotten methods to make things with real character and energy with a *human* component.

Therefore, to learn the most basic building blocks of a craft leads to mastery of all its aspects. We hope this book helps you along this path with its rich descriptions and illustrations!

About Hasluck's Traditional Skills Library

Paul N. Hasluck was a prominent author on artisan skills and traditional handcrafts toward the end of the 19th Century. He was the editor of the magazine *Work*, which was a popular handcraft, shop skills, and artisanal craft magazine of the day. His broad expertise in making things with your hands led him to write or edit over 30 volumes on specific handcrafts, arts, and mechanics, with each manual containing invaluable information related to each craft.

Hasluck had a great eye for collecting the info that beginners and experts alike needed to perfect their craft. His volumes were loaded with helpful diagrams, tables, and illustrations that are useful even by today's digital standards. In short, Hasluck's instructional manuals were the *go-to instructional library* if someone wanted to learn a particular skill. Used by the U.S. military, the Boy and Girl Scouts, and countless folks at farms, public libraries, and homes across the world, Hasluck's instructional manuals were the perfect "handy book" for learning.

This Doublebit Press Legacy Edition republishes this tradition of handcrafted quality and artisanal work. We hope that this deluxe printed edition of this work will help you gain mastery in your craft, as it is presented in the exact form that it was originally published. Even today, the knowledge contained within its pages are timeless and have much to teach!

Finally, as art, Hasluck's manuals contain beautiful illustrations and line art that are a sign of simpler, yet authentic times when quality mattered and craftsmanship was king. This collectible volume makes a great addition to the bookshelf of any handcrafter, maker, artisan, farmer, homesteader, or outdoors enthusiast!

PREFACE.

COLOURING MATTERS FOR DYEING TEXTILES contains, in a form convenient for everyday use, a comprehensive treatise on the subject. The contents of this manual are based on the highly esteemed book written by the late Dr. J. J. Hummel, F.C.S., Professor and Director of the Dyeing Department of the Yorkshire College, Leeds.

Without omitting any essential part of the original work the matter has been revised and brought up to date by Mr. A. R. Foster, Consulting Textile Expert, City and Guilds Honours Medallist. Needless to say many changes have taken place since the previous edition was published, and whilst the new processes and appliances have been incorporated, the older methods which are still in vogue in less progressive works, have been retained and revised. In this manner the manual has been made valuable, not only to the student, but all employed in bleaching, finishing, and dyeing works. It has, however, not been considered advisable to attempt to cope with all the new dyestuffs, as these are now numbered by thousands, and are being added to weekly. It has been deemed sufficient to give examples of each kind, as many of the new colours are simply mixtures.

Readers who may desire additional information respecting special details of the matters dealt with in this book, or instructions on any kindred subjects, should address a question to The Editor of WORK, La Belle Sauvage, E.C., so that it may be answered in the columns of that journal.

P. N. HASLUCK.

CONTENTS.

CHAPTER	PAGE
I.—Indigo Colouring Matters	9
II.—Logwood Colouring Matters	31
III.—Natural Red and Yellow Colouring Matters . .	47
IV.—Aniline Colouring Matters	50
V.—Quinoline and Phenol Colouring Matters . .	72
VI.—Azo Colouring Matters	84
VII.—Anthracene Colouring Matters	96
VIII.—Chrome Yellow, Iron Buff Manganese Brown, Prussian Blue	127
IX.—Method of Devising Experiments in Dyeing . .	132
X.—Estimation of the Value of Colouring Matters .	150
Index	155

LIST OF ILLUSTRATIONS.

FIG.		PAGE
1.	Indigo Grinding Mill	10
2.	Enlarged View of one Dyeing Cistern and Accessories	14
3.	Indigo Vat	18
4.	Apparatus for Preparing Hydrosulphite Vat Liquor	22
5.	Aniline Black Dyeing Machine	67
6.	Turkey-red Yarn-wringing Machine	98
7.	Tramping Machine for Turkey-red Yarn	100
8.	Clearing Boiler: Elevation	103
9.	Clearing Boiler: Plan	104
10.	Hydraulic Press	105
11.	Oil-padding Machine	108
12.	Liquor-padding Machine: Section	109
13.	Turkey-red Stove: Ground Plan	110
14.	Turkey-red Stove: Sectional Elevation	110
15.	Steaming-chest for Turkey-red Yarn	114
16.	Continuous Steaming-chest: Plan	116
17.	Continuous Steaming-chest: Elevation	116
18.	Experimental Dyeing Apparatus: Plan	140
19.	Experimental Dyeing Apparatus: Elevation	140
20.	Experimental Dyeing Apparatus: Section	140

CHAPTER I.

INDIGO COLOURING MATTERS.

Theory of Indigo Dyeing.—This valuable colouring matter is obtained from the leaves of various species of *Indigofera* (*I. tinctoria*, *I. disperma*, etc.), which are cultivated largely in India. It is also manufactured synthetically— whilst in addition to the artificial Indigo giving the usual blue shades and dyed in the same manner as natural Indigo, there is Thio Indigo Red B (Kalle) giving pink shades. The chief method employed in dyeing with Indigo is founded on the property it possesses of being converted under the influence of reducing agents (bodies capable of yielding nascent hydrogen) into indigo-white which is soluble in alkaline solutions. When textile materials are steeped for a short time in such solutions, and then exposed to the air, they become dyed blue in consequence of the re-oxidation of the indigo-white absorbed by the fibres, and the precipitation of insoluble indigotin thereupon, and, indeed, in such a manner as to be indelibly fixed. This "indigo-vat" method is applicable to all textile fibres, and gives permanent colours.

Another method of dyeing with Indigo, but one which yields fugitive colours, and is applicable only to the animal fibres, depends on the fact that Indigo treated with strong sulphuric acid becomes changed into soluble indigotin-di-sulphonic acid (Indigo Extract). Animal fibres attract and are dyed with this compound when simply steeped in its hot and slightly acidified solutions.

Vat-blue is largely employed, particularly in woollen dyeing, as the blue part of compound shades, as browns, drabs, etc. The same may be said of Indigo Extract or

10 FABRIC DYEING & TEXTILE COLORING MIXTURES.

Indigo Carmine blue, with regard to wool and silk dyeing. Indeed, this colouring matter possesses certain advantages over the majority of blue colouring matters. It can be associated with other acid colouring matters, and it dyes very level shades. Its only drawback is its extremely fugitive character.

Indigo Grinding Mills.—One of the first necessities in employing Indigo in dyeing is to have it thoroughly well ground. When required for making Indigo Carmine

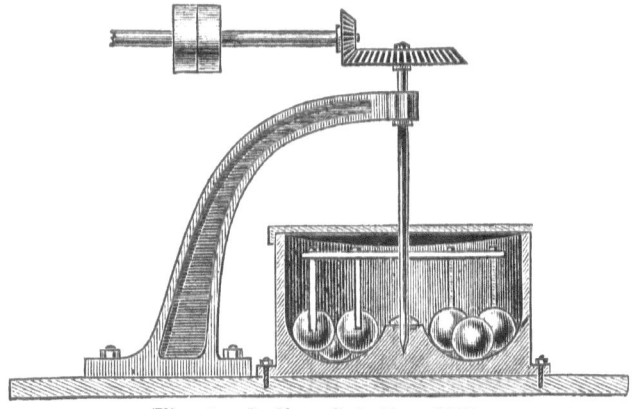

Fig. 1.—Indigo Grinding Mill.

it must be ground in the dry state, but for the indigo-vat it may be mixed with water, which considerably facilitates the grinding.

The oldest form of grinding machine is the stamping mill, provided with an arrangement for passing the ground Indigo through fine sieves. The mills now generally employed consist of cast-iron vessels, in which the Indigo is ground either by the rolling of heavy cannon balls or of iron cylinders. The ball mills are said to give the finest powder; the cylinder mills the greatest yield within a given period. Fig. 1 represents a section of one of the best forms of ball mills. It consists of a strong iron box enclosing several heavy cannon balls, which are pushed round by means of a pair of revolving arms. Sometimes the bottom of the box is flat, and heavy blocks of sand-stone are substituted for balls.

Application of Indigo to Cotton.—The fermentation

INDIGO COLOURING MATTERS.

vats so much used in dyeing wool with Indigo are never employed for cotton, since it is essential that this should be dyed in the cold to obtain the best colour; and, further, it permits the use of vats, in which the reduction of the Indigo is effected in a manner more under control. According to the reducing agents employed, the indigo-vats used for cotton may be named as follows: The ferrous sulphate vat, the zinc powder vat, the hydrosulphite vat.

Ferrous Sulphate Vat.—This vat, usually known as the lime and copperas vat, is the oldest and perhaps the one still most commonly employed. The vats or dye-vessels are rectangular tanks of wood, stone, or cast-iron. The size varies according to the material to be dyed; for calico they are generally 2 metres = 6 ft. 6 in. deep, 2 metres long, and about 1 metre broad, while for yarn-dyeing they are somewhat smaller.

In order to economise the Indigo as much as possible, the vats are generally worked in sets of ten.

The materials used in preparing this vat are:—

	Cloth.	Yarn.		Cloth.	Yarn.
Water	4,000 litres or	750 litres.	=	880·3 gals. or	165·0 gals.
Indigo	40 kilos. ,,	4 kilos.	=	88·1 lb. ,,	8·8 lb.
Ferrous sulphate 60-80	,,	6-8 ,,	=	132·2-176·3	,, 13·2-17·6
Slaked lime (dry) 50-100	,,	5-10 ,,	=	110·2-220·4	,, 11·0-22·0

The chemical changes which take place during the "setting" or preparation of the vat may be briefly summed up as follows: The lime decomposes the ferrous sulphate, and produces ferrous hydrate, which in the presence of the indigo rapidly decomposes the water, and becomes changed into ferric hydrate, while the liberated hydrogen at once combines with the indigotin to form indigo-white. This last substance combines with the excess of lime present, and at once enters into solution. These reactions may be expressed by the following chemical formulæ:—

$$FeSO_4 + Ca(OH)_2 = CaSO_4 + Fe(OH)_2.$$
Ferrous sulphate.   Lime.   Calcium sulphate.   Ferrous hydrate.

$$2[Fe(OH)_2] + 2H_2O = Fe_2(OH)_6 + H_2.$$
Ferrous hydrate.   Ferric hydrate.

$$C_{16}H_{10}N_2O_2 + H_2 = C_{16}H_{12}N_2O_2.$$
Indigotin.   Indigo-white.

The order in which the ingredients are added is of

comparatively little moment, and varies with different dyers. The most rational method, however, is to fill the vat with water and add first the ground Indigo and milk of lime; after raking up well, a solution of ferrous sulphate is added, and the whole mixture is systematically raked up at frequent intervals during twenty-four hours, until the Indigo is thoroughly reduced. With this plan the actual reducing agent, ferrous hydrate, is always in the presence of an excess of Indigo, and the indigo-white, the moment it is produced, is dissolved in the excess of lime. Owing to the mixture becoming rapidly thick and difficult to stir well, the more usual plan adopted is to put in the Indigo and ferrous sulphate first, and to add the milk of lime gradually. Lime is used in preference to caustic soda, because the vat thus produced dyes the cotton more readily; and owing to the film of calcium carbonate which forms on its surface, the indigo-white in the liquor beneath is less liable to become oxidised.

The ferrous sulphate employed should be as pure as possible. Any admixture of copper sulphate is injurious because of its oxidising influence, while the presence of aluminium sulphate and basic ferric sulphate, since these are quite inert as reducing agents, causes loss of so much lime as is required for their decomposition, besides a useless increase of sediment. The use of a large excess of ferrous sulphate and lime should also be avoided for this last reason. Ferrous sulphate, containing copper sulphate and ferric sulphate, is readily purified by boiling its solution with iron turnings, whereby the copper is precipitated, and the ferric sulphate is partly reduced to ferrous sulphate, or fully decomposed and precipitated.

A freshly made-up vat is in good condition when numerous thick dark blue veins appear on raking up the liquor, and the surface becomes rapidly covered with a substantial blue scum or "flurry." The liquid should be clear, and of a brownish-amber colour; if greenish, it shows the presence of unreduced Indigo, and requires a further addition of ferrous sulphate. If the colour is very dark, more lime is required.

At the end of every day's work the vats should be well raked up, and, according to their appearance, "fed" or replenished with small additions of lime and ferrous sulphate. The rake used for this purpose consists of a rectangular iron plate, with long wooden handle attached.

Before dyeing, the flurry should be carefully removed with an iron scoop or "skimmer," otherwise it attaches itself to the cotton, and causes it to look uneven or spotted.

Cotton yarn should be previously well boiled with water, in order to make it dye evenly. When dyeing light shades of blue, only a few hanks are dyed at once, the dipping, turning, and squeezing being performed with the utmost regularity. According to the depth of blue required, the duration of each immersion may vary from one to five minutes or more, and after wringing, the hanks are thrown aside, and allowed to oxidise completely.

The amount of indigotin which is precipitated on the cotton is said to vary with the duration of the immersion; if this be true it would appear that the cotton really attracts indigo-white from the vat solution, and is not dyed merely by reason of the indigotin precipitated from the portion of liquid absorbed by the fibre. The most economical method is to dye the cotton first in the weaker vats, and then to pass it through each succeeding stronger vat until the desired shade is obtained. For a dark shade the cotton should not be put at once into a strong vat, because it would be difficult in this way to obtain even colours; and in the long run the method would not be so economical. For light shades of blue only a few of the weaker vats are needed.

By this plan of always using the weakest vat first, so long as it yields any colour, each vat in turn becomes thoroughly exhausted.

After dyeing, the carbonate of lime which is deposited on the fibre is removed by rinsing in sulphuric acid, $2°$-$4°$ Tw. (Sp. Gr. 1·01-1·02). This operation removes the grey tint, and brightens the colour considerably. The cotton is finally dyed in a moderately strong vat, wrung out and dried at $60°$ C. This imparts to it the coppery lustre so much admired. It is, however, entirely superficial, and may be removed by simply washing in water.

As a rule, however, washing is avoided, since the Indigo is apt to rub off, and the colour may look bare and wanting in body and intensity. Vat blues are improved in colour by passing the goods through lime-water or a hot soap bath, probably because of the removal thereby of some yellow colouring matters.

A part of the Mather & Platt Indigo piece-dyeing machine is shown in Fig. 2, this portion being duplicated, trebled, or quadrupled according to the shade of indigo required. The cloth is taken through the vats in a continuous manner, passing over and under rollers beneath the surface of the liquor in the cistern A. After being squeezed by rollers B, it is laid by the winch C in loose

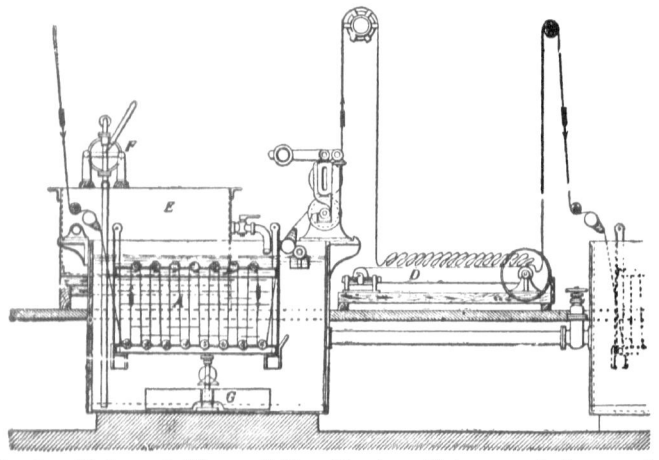

Fig. 2.—Enlarged View of One Dyeing Cistern and Accessories.

folds upon the slowly travelling apron D, the time of traverse being such as to enable the Indigo to become sufficiently oxidised upon the fibre. The cloth then enters the next vat, where the operations are repeated, and so on to the end of the range.

Each vat A is provided with a settling cistern E, having a hand pump F, whose suction pipe goes to the bottom of the dye-vat. By this means the deposit which forms in working can be removed before it has the opportunity to stain the cloth or cause uneven dyeing.

After leaving the last section, the pieces are first rinsed in cold water to remove the loose lime and Indigo adhering superficially, and then in dilute sulphuric acid, 4°-8°

INDIGO COLOURING MATTERS.

Tw. (Sp. Gr. 1·02-1·04), to dissolve off the calcium carbonate; they are finally washed and dried.

All the Indigo washed off in the rinsing pits, as well as the sediment of the vats themselves, must be collected in special tanks, in order to recover the Indigo.

According to F. C. Calvert, the vat sediments consist largely of an insoluble compound of indigotin and ferrous oxide, forming a bulky flocculent green precipitate. From this the indigotin can be recovered by decomposing it in the cold with strong hydrochloric acid.

Another method is to mix the vat sediments with water, and boil with some cheap, energetic reducing agent, such as caustic soda and orpiment. After settling, the clear liquid is drawn off, and oxidised by pumping it into a trough which stands at a high level, and allowing it to flow into a large tank; here the precipitated indigotin is washed and collected.

By adopting such methods of recovery, the total loss of Indigo may be reduced to 2-3 % of the original weight employed.

It is sometimes the custom to dye the cotton or "bottom" it with catechu brown, manganese bronze or brown, or a blue shade of aniline black, previous to introducing it into the Indigo vat. By this means very deep blue shades can be obtained with less Indigo than would otherwise be required.

It is well to remember that when aniline black is used the colour may be liable to become green on exposure.

In order to add a fictitious purple bloom or rich effect to the colour, vat blues are sometimes dyed afterwards or "topped" in a dilute solution of Methyl Violet or Methylene Blue, and dried without washing; less frequently they are dyed a logwood blue.

Zinc Powder Vat.—This vat is frequently used on the Continent, and also in Great Britain. It is founded on the fact that zinc, in the presence of lime and Indigo, readily decomposes water, and combines with its oxygen, whilst the liberated hydrogen reduces the indigotin to indigo-white, which is at once dissolved by the excess of lime present.

$$Zn + H_2O = ZnO + H_2.$$

The relative proportions used of the several ingredients

vary according to their quality, especially as regards the Indigo employed.

The following may be considered as average amounts:

Water	4,000 litres	=	880·3 gals.
Indigo	40 kilos.	=	88·1 lb.
Zinc powder	20	=	44·0 ,,
Slaked lime	20	=	44·0 ,,

The whole is well stirred occasionally during 18-24 hours, when it is ready for use. Lime and zinc powder are added as occasion requires. It is an extremely simple vat, easy to work, and possesses even certain advantages over the "lime and copperas" vat. In the first place, the sediment is reduced to about one-seventh of that in the vat referred to. Then the absence of ferrous sulphate removes the possibility of the formation of the insoluble compound of indigotin with ferrous oxide.

Hence this vat can be used without emptying for a much longer time than the "lime and copperas" vat, and there is little or no loss of Indigo.

Its chief defect is that it is liable to be muddy and frothy, from a continuous slight disengagement of hydrogen gas. Hydrogen is not given off until the whole of the Indigo is reduced, so that much froth denotes the presence of excess of zinc. If there be only little froth, it is removed by vigorously stirring up the vat several times, and then allowing it to settle, but with a large excess, a further addition of Indigo should be made before stirring. After settling for an hour, the vat should be sufficiently clear for dyeing.

If the vat is muddy, the same remedy must be applied, since the cause is the same. It is of no use to let it stand for a long time in the hope that it will settle; the hydrogen simply accumulates, and the liquid becomes still more muddy. The liquid must be vigorously stirred, in order to liberate the hydrogen from the sediment.

The dyeing should be completed before the liquid has had time to become muddy again. Experience alone can teach the exact amount of zinc powder which should be used, so that the vat may be maintained in an effective condition, yet free from the defects mentioned.

Some dyers find it an advantage to add about 12-20 kg. = 26·4-44·0 lb. of iron borings. These act mechanically, by presenting a large and rough surface, from which the hydrogen gas is more easily liberated, and thus a clear vat is more readily obtained.

Hydrosulphite Vat.—This vat is prepared for cotton exactly in the same way as for wool. The cotton, however, should be dyed in a cold solution.

Application of Indigo to Wool.—In order to utilise the Indigo to the fullest extent, it is previously ground with the addition of water, and added to the dye-vessel in the form of a fine smooth paste. The " vat " or dye-vessel in which the reduction of the Indigo and the dyeing takes place is a large tank, generally made of cast-iron (about 2 m. = 6 ft. 6 in. wide and 2 m. = 6 ft. 6 in. deep). For dyeing unspun wool it is generally round; for piece-dyeing, square. The whole is enclosed in brickwork, so arranged that the upper portion of the vat is surrounded by a chamber or canal, into which steam can be admitted. By this means the liquid of the vat is heated from the outside, and a regular temperature can always be maintained, without danger of disturbing the sediment.

During the "setting" of the vat the contents are stirred up, either by hand, by means of a rake, or by a mechanical arrangement fixed in the bottom of the vat, and driven by machinery. Before dyeing, the contents are allowed to settle, since the textile material must always be dyed in the clear liquid. The disturbing of the sediment is prevented as much as possible by suspending in the vat, 1 m. = 39 in. below the surface, a so-called "trammel," *i.e.* an iron ring or frame, across which coarse rope network is stretched.

Fig. 3 gives the section of a well-arranged round indigo-vat for wool dyeing, with mechanical stirrer, etc. A is the steam-chamber surrounding the vat, B the steam-pipe for heating it, J the trammel-net, I the emptying-pipe; D is a fixed bar supporting the stirring-screw C and the cone E. The parts drawn in dotted lines represent the movable portions of the stirring arrangement; G is a strong wooden bar which can be readily fixed across the top of the vat; it supports a pair of cog-wheels, and the fast and loose pulleys H; F is a vertical connecting shaft, which, by reason of the guiding cone E, can be readily connected with the screw C. A cheaper but less substantial arrangement is that in which the vat is made of wood, and heated internally by a copper steam-pipe forming a spiral half-way up the walls of the vat. When not in use for dyeing, and to prevent loss of heat and oxidation of the reduced indigo, the vat is covered with a wooden lid, which is divided into two or three pieces, for the sake of convenient handling. It is often the custom to throw over it a woollen cloth cover in addition. According to the materials used in preparing or " setting "

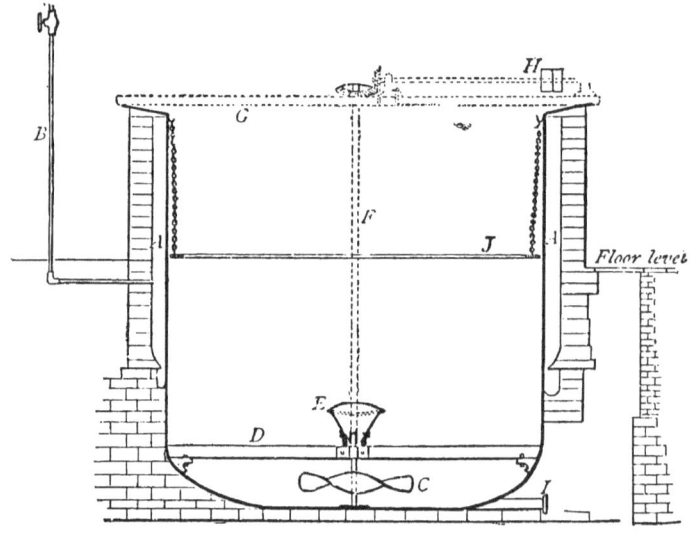

Fig. 3.—Indigo Vat.

a vat for woollen dyeing, the following vats may be distinguished: Woad, potash, soda, urine, and hydrosulphite.

Woad Vat.—In setting this vat the following substances are employed for a vessel of the dimensions already given:

Indigo	15 kilos.	=	33·0 lb.
Woad	300	=	661·3 ,,
Bran	10	=	22·0 ,,
Madder	2-15	=	4·4-33·0
Dry slaked lime	12	=	26·4 ,,

The vat is first partly filled with water, the crushed woad is then added, and the whole is well stirred up and heated to about 50°-60° C. This temperature is maintained for 24-30 hours, the stirring being repeated at intervals during the first 2 hours. The well-ground Indigo, also, the bran, madder, and about half the total quantity of lime, are now added; after well raking up the whole mixture, the vat is covered over, and left to itself for 12-24 hours.

By this time fermentation has generally well begun. It is recognised by the following appearances The surface of the vat-liquor becomes covered with a coppery-blue scum or "flurry." On gently stirring the liquor, it is seen to possess a greenish-yellow colour, interspersed with blue veins or streaks of regenerated Indigo, and the general odour of the vat is agreeable. If the bottom of the vat-liquor is disturbed, a slight froth appears on the surface, and on bringing up a portion of the sediment with the rake, it shows evidences of being in a state of slight fermentation, and smells somewhat sour. A piece of wool immersed in the liquid for a short time, and then exposed to the air, becomes dyed blue. All these appearances denote that the fermentation is progressing satisfactorily, and it now only becomes necessary to keep it steady and under control, by maintaining the temperature at 45°-50° C., adding, every two or three hours, a portion (one-eighth to one-fourth) of the remaining quantity of lime, and vigorously stirring the whole contents of the vat after each addition. In the course of the next 12-24 hours, provided the fermentation continues to progress favourably, the vat is ready to be used for dyeing. Excessive fermentation is prevented by well-timed and suitable additions of lime; sluggish fermentation, on the contrary, is accelerated by making further additions of bran. The dyeing power of the vat is maintained by adding, after each day's work, fresh quantities of lime and bran, and every other day 5-8 kg. = 11·0-17·6 lb. Indigo: care being taken to keep the temperature of the liquor at about 50° C. After three or four months, or whenever the vat sediment becomes so bulky that there is a difficulty in obtaining the clear-liquor space necessary for good dyeing, no further additions of Indigo are made; the vat is then used for dyeing light blues, and when its colouring power is exhausted, the whole contents are

20 FABRIC DYEING & TEXTILE COLORING MIXTURES.

thrown away. The woad vat gives rich and brilliant colours, and serves equally well for light and dark shades of blue. It is the vat most largely employed in Yorkshire for woollen dyeing.

Potash Vat.—This vat is made up as follows:—

Indigo	10 kilos. =	22·0 lb.
Madder	2·5	= 4·4-11·0
Bran	2·5	= 4·4-11·0
Carbonate of potash	10-15	= 22·0-33·0

The bran and madder are first heated to 80°-100° C. for 3-4 hours with water, after which the potash is added and dissolved, and the liquor is allowed to cool down to about 40° C. The ground Indigo is then added, the whole is well stirred, and left for a period of 48 hours to ferment, an occasional stirring every 12 hours or so being needed.

The appearances of a healthy state of fermentation in the potash vat are similar to those observed in the woad vat.

This vat, owing to the absence of such a highly nitrogenous substance as woad, is less liable than the woad vat to get out of order, and is altogether more easily managed. It also dyes more rapidly than the woad vat, gives deeper but duller shades of blue, and the colour does not come off so much on milling with soap and weak alkalis. It is best adapted for very dark shades of navy blue. If unspun wool is dyed in this vat, care must be taken to wash it thoroughly in water afterwards, otherwise it is apt to spin badly.

Soda Vat (German Vat).—This vat is set with the following materials: Indigo, 10 kg. = 22·0 lb.; bran, 60-100 kg. = 132·2-220·4 lb. (or 10-15 kg. = 22·0-33·0 lb. of treacle instead); carbonate of soda crystals, 20 kg. = 44·0 lb.; slaked lime, 5 kg. = 11·0 lb.

The bran is first boiled with the water for 2-3 hours, the liquid is then cooled down to 40°-50° C., the remaining ingredients are added, and the whole is well stirred up and left to ferment for 2-3 days, with only an occasional stirring. During the progress of the fermentation, lime and soda, as occasion requires, are added from time to time. After being used for dyeing, the vat is replenished with Indigo, soda, and lime. This vat is cheaper than the potash vat, because of the difference in price between potash and soda; it also lasts longer. It is, however, more liable to get out of order, though always more easily managed than the woad vat.

Urine Vat.—This vat, although of minor importance, is suitable for working on a small scale. It is used by those who only require to dye vat-indigo blues occasionally, or in comparatively small quantities. The vat is made up as follows Stale urine, 500 litres = 110 gals.; common salt, 3-4 kg. = 6·6-8·8 lb. Heat the mixture to 50°-60° C. for 4-5 hours, with frequent stirring; then add 1 kg. of madder and 1 kg. of ground Indigo, stir well, and allow to ferment till the Indigo is reduced. In this vat the indigo-white dissolves in the ammonium carbonate arising from the decomposition of the urea contained in the urine.

Hydrosulphite Vat.—The active reducing agent in this vat is a solution of hyposulphurous (hydrosulphurous) acid, which may be produced by the action of zinc upon a solution of sulphurous acid, according to the following equation:—

$$H_2SO_3 + Zn = H_2SO_2 + ZnO.$$
Sulphurous acid.   Zinc.   Hyposulphurous acid.   Zinc oxide.

In practice the zinc is allowed to act upon a concentrated solution of sodium hydrogen sulphite (bisulphite), instead of sulphurous acid, in which case the reaction is somewhat more complicated, there being produced a solution of sodium hydrogen hyposulphite and zinc sodium sulphite which separates out, thus:

$$Zn + 3NaHSO_3 = NaHSO_2 + ZnNa_2(SO_3)_2 + H_2O.$$
Sodium hydrogen sulphite.   Sodium hydrogen hyposulphite.   Zinc sodium sulphite.

The reduction of the indigotin by means of the acid sodium hyposulphite may be represented by the following equation:—

$$C_{16}H_{10}N_2O_2 + NaHSO_2 + NaHO =$$
Indigotin.   Acid sodium hyposulphite.

$$C_{16}H_{12}N_2O_2 + Na_2SO_3.$$
Indigo-white.   Disodium sulphite.

It is not customary to reduce the Indigo in each vat separately, but rather to make a very concentrated solution of reduced Indigo, and to use this stock solution for preparing and replenishing the dye-vats.

The setting of a hydrosulphite vat naturally divides itself into three phases:—

1. The formation of acid hyposulphite of soda according to the above equation.

2. The changing of this acid hyposulphite into neutral hyposulphite by mixing it with lime.

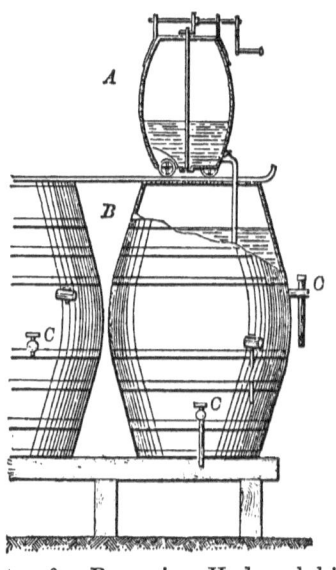

Fig. 4.—Apparatus for Preparing Hydrosulphite Vat Liquor.

3. The mixing of this solution with Indigo and a further quantity of lime, in order to produce the stock solution of reduced Indigo. Fig. 4 shows a suitable apparatus in which to conduct the first two operations.

(1.) A vessel A, provided with an agitator, and which can be hermetically closed, is packed full of small rolls of zinc-foil, then filled up with bisulphite of soda, 55° Tw. (Sp. Gr. 1·275), and thoroughly saturated with sulphurous acid. The size of the vessel should be just adapted to the quantity of hyposulphite required for immediate use, and so that it may be entirely full of zinc, in order to prevent oxidation as much as possible. The zinc and bisulphite of soda are allowed to act upon each other, with occasional stirring, for about an hour at least. The sodium hydrogen hyposulphite thus produced stands at 62° Tw. (Sp. Gr. 1·31), and, since it is very unstable, it must be used at once, either for reducing Indigo or for making the neutral sodium hyposulphite. One litre = 1 quart bisulphite of soda, 55° Tw. (Sp. Gr. 1·275) requires 100-125 g. of zinc, of which quantity about 50 g. = 1·6 oz. dissolve during the operation. Granulated zinc or zinc powder may be used instead of zinc-foil, but the former retains much of the liquid when the vessel is emptied, and cannot be so readily washed. The latter is perhaps the best form of zinc to use, but it varies considerably in composition, and owing to its being in the state of such a fine powder, the liquid heats considerably during the mixing, so that there is always the danger of a portion of the hyposulphite being decomposed. Whenever the acid hyposulphite of soda has been drawn off, the zinc should be rinsed with water, and, if not immediately required again, the vessel should be filled up with water, in order to prevent, as much as possible, the oxidation of the zinc.

When a fresh quantity of hyposulphite is to be made this water is drawn off and the zinc is rinsed, first, with water slightly acidulated with hydrochloric acid, and afterwards with water. The small quantity of zinc dissolved in the previous operation is replaced by an addition of fresh zinc-foil, so that the vessel may be entirely full. Bisulphite of soda solution is then poured over it, and the process as already described is repeated.

(2.) In order to change the unstable acid sodium hyposulphite thus produced into the more stable neutral hyposulphite, it is drawn off into another closed vessel B, and there mixed with milk of lime, which precipitates zinc oxide and calcium sulphite. One litre = 1 quart of acid sodium hyposulphite, 62° Tw. (Sp. Gr. 1·31) requires about 460 g. = 14·7 oz. of milk of lime, containing 200 g. = 6·4 oz. of quicklime per litre = 1 quart. The mixture is well agitated, and, after settling, the clear liquid is drawn off. A better yield is obtained by passing the

24 FABRIC DYEING & TEXTILE COLORING MIXTURES.

mixture through a filter-press. The weight of neutral hyposulphite thus obtained is about the same as that of the original sodium bisulphite, and it has a density of about 36° Tw. (Sp. Gr. 1·12). It is best to employ the solution as soon as possible for reducing Indigo, and not to make more than is required for immediate use. If it is ever found necessary to keep it for some time, it must be made alkaline by adding a little lime, but this should only be done as an exception and not become general.

(3.) The stock solution of reduced Indigo is made by heating to a temperature of 70°-75° C. the following mixture : Indigo, 1 kg. = 2·2 lb.; milk of lime (containing 200 g. 6·4 oz. of quicklime per litre = 1 quart of water), 1-1·3 kg. = 2·2-3 lb.; and so much neutral hyposulphite, 36° Tw. (Sp. Gr. 1·18), as is obtained from 8-10 kg. = 17·6-22·0 lb. of concentrated sodium bisulphite. The Indigo is rapidly and completely reduced, and a comparatively clear greenish-yellow solution is obtained, containing about 1 kg. of Indigo per 10-15 litres = 2·-3·3 gals. of solution. With an insufficiency of lime, part of the indigo-white is not dissolved, but remains as a dense white precipitate. In setting a hydrosulphite vat, the vat is first filled with water heated to 50° C.; it is then deprived of the oxygen it naturally contained, by adding a little of the neutral hyposulphite. The concentrated stock solution of reduced Indigo is then added in sufficient quantity to make a vat of the required strength, and since there is no sediment, the dyeing may be at once proceeded with. The dyeing power is maintained by adding fresh quantities of the concentrated solution of reduced Indigo. The liquid of the vat should always contain an excess of hyposulphite. It should have a yellow colour, and be clear.

If from any cause excessive oxidation of the indigo-white takes place, and the liquor becomes greenish, a little more hyposulphite, and possibly also milk of lime, must be added, and the whole heated to 70°-75° C., in order to accelerate the reduction of the Indigo and restore the normal yellow colour. When the vat is in use the alkalinity of the liquid increases, and there is a danger of both the colour and the fibre being injured; hence it is advisable to partially neutralise the excess of alkali from

INDIGO COLOURING MATTERS.

time to time, by making slight additions of dilute hydrochloric acid.

Defects in Indigo Vats.—All the fermentation vats are subject to derangements, by which they become more or less useless. The most serious defect is produced by using a deficiency of lime, in which case the fermentation becomes more and more active; if allowed to proceed too far, the Indigo is totally and irretrievably destroyed.

This defect is recognised by the following characteristics : The flurry disappears, the vat liquor has a muddy appearance, and gives off a very disagreeable odour; it has a dirty reddish-yellow tint, and acquires the property of gradually destroying the colour of a small piece of indigo-blue cloth which may be plunged into it.

The only remedy to be applied, in such a case, is to heat the vat liquor to 90° C., and add lime or potash, etc., according to the kind of vat. If this has not the effect of arresting the fermentation, the vat will " run away," or be " lost," as the dyers term it. The woad vat is the one most liable to this defect, because it contains a very large quantity of nitrogenous matter.

Another defect is caused by dyeing too many pieces in rapid succession. Under these circumstances the appearance of the vat becomes very similar to that exhibited in the last defect, but the odour is faintly ammoniacal. A vat in this state gives much paler colours than when it is in its normal condition. To remedy this defect, it is advisable, first, to add a little lime—perchance one might have been deceived by the colour of the vat—and, after some time, to add a little woad and warm the vat, endeavouring thus to promote rapid reduction of the precipitated Indigo, and to get rid of the excessive amount of oxygen which has been introduced into the vat. A quicker method is to add a small quantity of ferrous sulphate, and then to stir up the liquor thoroughly.

A third defect is caused by the presence in the vat of such an excess of lime as to precipitate the indigo-white. In this case, the vat liquor becomes of a dark brown colour, and the healthy odour and the blue scum or " flurry " disappear. When noticed in time, this defect is remedied by adding at intervals a small quantity (say ½ kg. = 1 lb.) of ferrous sulphate, or of dilute sulphuric acid, and stirring up the vat. This addition precipitates the excess of lime as calcium sulphate.

Dyeing with the Indigo Vat.—The operations involved in setting, working, and keeping the fermentation indigo-vats in order are very simple, but since the phenomena of fermentation are for the most part extremely complex and ill-understood, long-continued and close observation is necessary before one is able at a glance to recognise the features characteristic of the very various conditions which the vat liquid may assume.

Woollen material, of whatever kind, must always be well boiled with water, and then at once passed into tepid water and squeezed before it is entered into the indigo-vat. It should never be allowed to lie in irregular heaps after boiling, since this causes unequal dyeing. This wetting-out process not only accelerates the absorption of the vat liquor by the material, but also produces more even dyeing, and prevents the introduction into the vat of a large amount of air, which would oxidise the indigo-white, and cause a precipitate of indigo-blue merely to adhere mechanically and loosely to the surface of the material. Cloth previously milled with soap should be well boiled with water and washed, in order to remove all traces of soap, otherwise " cloudy " or irregular dyeing will result, through the precipitation on the cloth of a lime-soap, which, being of a sticky nature, acts as a resist. Before beginning to dye, the blue scum on the surface of the vat liquor must always be skimmed off, in order to prevent it from producing irregular spots on the material, especially in the case of cloth-dyeing.

Loose wool is gently moved about, by means of poles, in that portion of the vat liquor which is above the trammel-net, care being taken at this stage never to bring it above the surface and thus expose it to the air. Loose wool is best dyed in a vat in which the fermentation is slightly active. When it has been immersed, and worked the length of time necessary to obtain the intensity of colour required (say ten minutes to two hours), it is taken out, placed in strong bags of netted cord, and the excess of liquor is well wrung out. The wool is then thrown into heaps, and remains exposed to the air until the blue colour is fully developed. The dyed wool must

INDIGO COLOURING MATTERS.

now be well washed with water slightly acidulated with sulphuric or hydrochloric acid, in order to remove loosely adhering Indigo and all soluble material absorbed from the vat, also to dissolve away any adhering lime carbonate. This treatment prevents the greyish appearance the blue would otherwise acquire on drying, and gives it greater brilliancy. The acid must, of course, be entirely removed by a final washing in water before drying.

In dyeing woollen yarn each hank is worked in the vat separately for a short time, and at once wrung out and thrown on the floor to oxidise. To obtain dark shades the whole process must be repeated several times. The subsequent operations of washing, etc., are the same as in the case of dyeing loose wool.

Woollen cloth, after boiling and cooling as already mentioned, is dyed by moving it about, or "hawking" it, in the vat liquid above the trammel-net, by means of an instrument called a "hawk," a double hook, one of which is held in each hand by the workman. Care must be taken during the whole operation not to raise the cloth above the surface of the liquid, in order to avoid irregular oxidation, and, consequently, uneven dyeing. The duration of the hawking process may vary from twenty minutes to two hours, according to the dyeing power or strength of the vat, the texture of the cloth, and the depth of colour required. Since the vat solution is not so readily and thoroughly absorbed by thick and closely woven cloth, such material requires longer time than if it were thin and loosely woven. In many well-ordered dye-houses the moving about of the cloth in the vat liquor is effected by a so-called hawking machine. This consists essentially of a framework, supporting a pair of squeezing rollers a little below the surface of the vat liquor. The cloth to be dyed is opened out, and the ends are stitched together, so as to form a broad, endless band; this is continually drawn through the vat liquor and between the squeezing rollers until the shade required is obtained; the squeezing rollers are provided with close-fitting iron scrapers, which prevent the cloth from wrapping round them, and the framework has guiding pegs to keep the cloth from running to one side. The squeezing rollers are turned either by hand or steam-power. Such machines give much more regular work, and in thick cloth the dye may be made to penetrate more thoroughly to the centre of the fabric, it being possible to regulate at will the pressure of the squeezing rollers.

In order to dye to an exact shade of dark blue, the cloth is worked in a strong vat until a blue a little lighter in shade than that ultimately required is obtained; it is then withdrawn from the vat, and the colour is fully developed by exposure to the air, after which the operations are repeated in a weaker vat till the desired shade is gained. If a light shade of blue is wanted, the cloth is at once worked in a weak vat for the requisite length of time. Cloth dyes best in a vat in which the fermentation is kept very moderate. After dyeing, the cloth must be rinsed in acidulated water, and well washed, as already mentioned in the case of loose wool. In order to remove every trace of loosely adhering Indigo, a good milling with soap and fuller's earth is eventually given, so that the finished piece will not soil a white handkerchief when rubbed on its surface. Previous to fulling it is well to boil the goods with a solution of alum or bichromate of potash and tartaric acid. This operation makes the colour much faster against light and rubbing. As a rule, the loose Indigo removed during the fulling operation is allowed to run to waste, but it is probable that much of it might be profitably regained from the sediment of the fulling mills by a system of washing based on the difference of specific gravity between the fuller's earth and indigotin; or one might mix the sediment with ferrous sulphate solution, let settle, draw off the clear solution of reduced Indigo, oxidise it, and collect the regenerated and precipitated Indigo.

By boiling the cloth (after dyeing and washing) with Barwood, Sanderswood, or Camwood, the blue is said to be better fixed and faster against light than without this treatment. It is not so liable to bleach at the cut edges, etc. Steaming the goods for half an hour makes the blue a little more violet, but faster against light. Vat-blues, which have been boiled with alum and tartar after dyeing, become slightly paler by the operation, but are rendered much more stable against light; the ultimate gain being much greater than the primary loss,

INDIGO COLOURING MATTERS.

Exposure to light gives vat-blue a violet tint. By boiling with alum and tartar, after dyeing and washing, etc., and steaming half an hour in addition, the colour is only slightly weaker, and the greatest fastness against light is obtained. If treated in this way, the colour becomes darker during the first five months of exposure to light and air, and at the end of the year possesses the same depth of shade as at first. From these facts it may be inferred that in the case of woaded browns, greens, etc., the dyeing in the woad-vat should precede the mordanting and dyeing of the added colours.

Silk is now seldom dyed vat-blue. Lime tends to make the silk harsh and brittle, hence it is well to employ the soda or potash vat, the hydrosulphite vat, or, better still, a vat in which the Indigo is reduced by zinc powder and ammonia.

Indigo Extract; Indigo Carmine $[C_{16}H_8N_2O_2(HSO_3)_2]$. —This colouring matter is the product of the action of strong sulphuric acid on Indigo.

Indigo Extract has no affinity for cotton, and cannot be used by the cotton dyer. It is, however, occasionally used by bleachers for tinting.

Application of Indigo Extract to Wool.—The blue obtained on wool by means of Indigo Extract and Indigo Carmine is sometimes called " Saxony blue." It is a much brighter colour than that obtained by means of the indigo-vat, but is very far from being so fast either to light or to the action of soap and weak alkalis. Hence it does not stand milling well. Wool must always be dyed with the above-mentioned colouring matters in an acid bath. When " sour extract " (*i.e.* indigotin-disulphonic acid, containing free sulphuric acid) is used, no other addition to the dye-bath than the extract itself is necessary. The wool may be entered at 40°-50° C.; the temperature of the bath should then be gradually raised in the course of half an hour to the boiling point, the dyeing continued for half an hour longer. By dyeing at 70°-80° C. a purer blue is obtained, but the colour is apt to have an uneven, speckled appearance. Boiling levels the colour, but makes the shade greener. When Indigo Carmine is employed, this being the sodium salt of indigotin-disulphonic acid, it is necessary to add to the dye-bath, along with the colouring matter, 5-10 % of sulphuric acid, 168° Tw., so that this may combine with the sodium and liberate the colour-acid itself. Without this addition the full colouring power of the Indigo Carmine would not be developed. The addition of 10-20 % of sodium sulphato to the dye-bath along with sulphuric acid tends to make the colour uniform or level. Sometimes alum is also added to the bath in order to mordant the wool slightly, and permit the application of Logwood and other polygenetic colouring matters.

Application of Indigo Extract to Silk.—Dye at a temperature of 40°-50° C. in a bath acidified with sulphuric acid and containing the amount of Indigo Carmine solution necessary to produce the depth of shade required.

Another method is to mordant the silk first with alum by steeping twelve hours in a solution of 25 % of alum, and then, without washing, to dye in a solution of Indigo Carmine with the addition of about 10 % of alum to the dye-bath. If scroop is required, a further addition of a little acetic acid or cream of tartar is necessary.

In this case the alum acts in no sense as a mordant for the Indigo Carmine, but makes it possible to redden the shade, or even to produce a violet colour, by adding Cochineal decoction to the dye-bath; by the further addition of decoctions of Old Fustic, Logwood, Orchil, etc., various shades—grey, drab, brown, etc.—may be obtained, according to the amount of each colouring matter employed.

By adding to the dye-bath a decoction of 10-20 % of Logwood a dark shade of blue is obtained; the addition of too much Logwood decoction, however, must be avoided, otherwise the colour is apt to become dull. The most rational method of adding the colour yielded by Logwood to that of the Indigo Carmine is to dye with the two colouring matters in separate baths.

CHAPTER II.

LOGWOOD COLOURING MATTERS.

Logwood.—This dyewood is the heart wood of *Hæmatoxylon campechianum*, grown in Central America.

Application of Logwood to Cotton.—The principal use of Logwood in cotton-dyeing is for the production of blacks and greys; it may, however, also serve for purples, blues, and numerous composite colours. In conjunction with other colouring matters, it is employed for the production of numerous compound shades, its use being, in such cases, to make the colour darker, or of a bluer tone.

Logwood Blacks.—The method of obtaining a logwood black consists essentially in mordanting the cotton with a salt of iron, and then dyeing with a decoction of Logwood. Numerous modes of applying this simple process are in general use, but the principle is always the same.

In order to mordant the cotton, it may be worked in a cold solution of pyrolignite or nitrate of iron, at about 5° Tw. (Sp. Gr. 1·025) till thoroughly saturated; after squeezing, the iron is fixed by working in a cold weak bath of sodium carbonate, or milk of lime; the cotton is finally well washed in water.

Another method of mordanting, and one which gives faster blacks, is to fix on the fibre a tannate of iron instead of ferric oxide, as in the last case. Work the cotton in a cold infusion of about 30-40 % of Sumach, or its equivalent of other tannin matter (ground Gall-nuts, Myrabolams, etc.), and allow it to steep for several hours, or even over-night; remove the excess, and, without washing, work for about half an hour in a cold solution of pyrolignite or nitrate of iron at 2°-4° Tw. (Sp. Gr. 1·01-1·02), and wash well. In order to remove all traces of acid, and to fix more completely on the fibre a basic salt of iron, it is advisable before washing to work the cotton in a cold bath of chalk-water, or in weak milk-of-lime. Not unfrequently a lime bath is applied immediately after sumaching and before passing into the iron bath. In this case a tannate of lime will be formed upon the fibre, and the double decomposition with the iron salt is facilitated, since the lime at once takes up the acid liberated.

In warp dyeing the whole process is continuous, and the cotton, after being steeped in a decoction of myrabolams, is passed successively through baths containing lime-water, nitrate of iron, logwood liquor, dilute iron solution, and water.

For low-class goods many dyers substitute ferrous sulphate for the pyrolignite and nitrate of iron.

The pyrolignite of iron may also be mixed with an equal or somewhat smaller amount of aluminium acetate (red liquor) at 5° Tw. (Sp. Gr. 1·025), in which case it may be better to fix the mordants by working the cotton for a quarter of an hour at 50°-60° C., in a dilute solution of phosphate or arsenate of soda.

An aluminium mordant alone would give a dull lilac shade, but along with an iron mordant it helps to remove the unpleasant reddish or rusty appearance of the blacks otherwise obtained.

When Catechu is the tannin matter employed, the cotton should be worked in a boiling decoction of it, and allowed to steep till cold, in order to effect the precipitation on the fibre of the maximum amount of catechin. The cotton may afterwards be worked 5-15 minutes in a boiling solution of bichromate of potash (5 g. = 0·17 ozs. per litre = 1 quart), before passing it into the bath of pyrolignite of iron, though this is not absolutely necessary.

By whichever method the mordanting is effected, the dyeing takes place in a separate bath containing a suitable amount of freshly-made Logwood decoction, together with a small quantity of extract of Old Fustic, or of Quercitron Bark. If an iron mordant only has been employed, it is beneficial to add also a small quantity of copper sulphate to the dye-bath, in order to prevent the cotton from acquiring the rusty appearance already referred to.

The cotton is introduced into the cold dye liquor, and the temperature is gradually raised to the boiling point.

LOGWOOD COLOURING MATTERS.

After dyeing, the cotton may be passed through a solution of bichromate of potash, 0·5 g. = 0·017 oz. per litre, at 60° C. This operation gives intensity and fastness to the black, since any excess of colouring matter is fixed as a chromic oxide lake.

The dyed cotton is washed and worked in a solution of soap, 5 g. = 0·17 oz. per litre (1 quart), at a moderate temperature, then squeezed and dried. This final soaping removes any bronze appearance, and imparts to the colour a bluer and more agreeable tone. The cotton also acquires a softer feel.

The following is a method by which a chrome black on cotton can be obtained in a single bath

Dissolve 1·5 kg. = 3·3 lb. of bichromate of potash in a small quantity of water, mix the solution with 500 litres = 110·0 gals. of Logwood decoction at 3° Tw. (Sp. Gr. 1·015), and add 3·5 kg. = 7·7 lb. of hydrochloric acid, 34° Tw. (Sp. Gr. 1·17). The cotton is introduced into the cold solution, and the temperature is very gradually raised to the boiling point. The cotton acquires at first a deep indigo-blue shade, which changes to a blue-black on washing with a calcareous water.

A slight modification of this process which may be adopted is to work the cotton in a solution containing at first only the bichromate of potash and hydrochloric acid, and to add the decoction of Logwood to the bath, in small portions from time to time, gradually raising the temperature as before.

Another method of producing a Logwood black is to use a bath containing Logwood extract and copper acetate, entering the cotton cold, raising the temperature gradually to 50° C., and dyeing at that temperature until the colour is sufficiently developed.

Copper sulphate, to the amount of about 4 % of the weight of cotton, is frequently used instead of acetate, and an addition of 4 % of soda-ash is made to the bath along with 20 % of solid Logwood extract.

The cotton is passed rapidly through this mixture, heated to 60°-80° C., and then allowed to oxidise or " smother " for 5-6 hours. This process requires to be repeated several times before a full black is obtained. The method is not economical for ordinary use, but it is said to yield a black which withstands milling with soap very well. Carbonate of copper may also be applied in the above process, instead of copper sulphate and soda-ash.

The methods already given may be adapted to the dyeing of unspun cotton. The following method of dyeing a chrome black is said to be specially applicable to such as must withstand the operation of fulling. Wet out the cotton well in boiling water, then boil in a strong solution of about 30 % of solid Logwood extract, drain, and allow it to lie exposed to the air for some time; complete the oxidation thus begun by working it one hour in a cold solution of 8 % of bichromate of potash and 6 % of copper sulphate, wash and complete the dyeing in a bath containing 10 % of Logwood extract; enter the cotton cold, and raise the temperature gradually to the boiling point. Wash, soap, and dry.

In the first bath the cotton simply absorbs the colouring matter of the Logwood; in the second this is oxidised, and at the same time combined with a sufficient amount of mordant, copper, and chromic oxide, to enable it to take up still more colouring matter in the third bath. The first Logwood bath is analogous to the tannin bath alluded to in a previous process (p. 31).

Logwood Greys are obtained by working the cotton for a short time at 40°-50° C. in a weak decoction of Logwood (1-5 %), then in a separate bath containing a weak solution of ferrous sulphate or potassium dichromate, and washing. Many dyers adopt the apparently irrational method of mixing the ferrous sulphate and Logwood solutions, and dye at once in the inky liquid thus obtained. Comparatively little precipitate, however, is produced in the dye-bath in this case, and the colour is, for the most part, developed on the cloth itself during the subsequent oxidation by exposure and washing. The shade of grey may be modified *ad libitum* by adding to the Logwood bath a small proportion of decoctions or extracts of tannin matter, Old Fustic, Peachwood, etc.

Logwood Purples are obtained by mordanting the cotton in a weak solution of stannous chloride, then washing and dyeing in a separate Logwood bath. The colour is tolerably fast to soap, but not to light.

Logwood Blues on cotton are now seldom dyed, because of their fugitive character. To obtain them, work the cotton in a bath containing a decoction of Logwood and a small proportion of copper acetate or sulphate, raising the temperature gradually to 50° C. The tone of colour has great similarity with that of an indigo-vat blue.

Application of Logwood to Wool.—Logwood is the essential basis of all good blacks on wool, although other colouring matters are frequently used along with it, either to modify the particular shade of black, or to add to its intensity and permanence.

According to the materials employed, we may distinguish the following kinds: Chrome black, copperas black, and woaded black.

Chrome Blacks are produced by first mordanting the wool for 1-1½ hour, at 100° C., with 3 % of bichromate of potash and 1 % of sulphuric acid, 168° Tw. (Sp. Gr. 1·84), then washing and dyeing in a separate bath for 1-1½ hour, at 100° C., with 35-50 % of Logwood. This represents the simplest form of dyeing a chrome black, but in practice numerous slight modifications are introduced, in order to obtain various shades of black. The following, which are typical, may be mentioned. The mode just given yields a blue-black, or, as it is sometimes called, a black with blue reflection. By the addition of a suitable amount of some yellow colouring matter to the dye-bath—say, 5 % Old Fustic—a dead-black is obtained, that is, a neutral black, which possesses no decided tint of blue, green, violet, etc. By increasing the amount of Old Fustic to 10 %, a green-black is obtained, and the greenish shade becomes still more pronounced if 3-4 % of alum is added to the mordanting bath along with the bichromate of potash.

A violet-black is produced by dyeing exactly as for blue-black, but after the dye-bath has been exhausted, a dilute solution of about 2 % of stannous chloride (tin crystals), or its equivalent of commercial muriate of tin containing no free acid, is added to the dye-bath, and the boiling is continued 15-20 minutes longer.

In the case of dead-blacks it is the custom with some dyers to "sadden" in a similar way with 3-4 % of ferrous sulphate; or, instead of this, the goods are passed, after dyeing, through a warm bath containing about 0·5 % of bichromate of potash. The object of these last modifications is to precipitate and fix more completely on the wool any colouring matter, perchance not combined with the mordant, but simply absorbed by the wool.

With black yarn, which will eventually appear in a woven fabric, in close proximity to white or delicately-coloured yarns, this fixing of the dye is very necessary, otherwise the light-coloured or white yarns become stained during the operations of milling, etc., and the finished fabric has a soiled appearance. It is always the case that some black comes off during these operations, but if the colouring matter of the Logwood is thoroughly combined with its own mordant, it will not readily combine with the mordant of any neighbouring fibre, but be simply rubbed or washed out as an insoluble powder.

Chrome blacks may also be dyed in a single bath, as follows : A mixture of Logwood liquor and bichromate of potash solution in suitable proportions is boiled. The precipitate thus produced is collected, and may then be employed as a " direct black " or a " one-dip dye." It is, indeed, the actual coloured body or pigment one wishes to fix upon the wool, and this is rendered possible, because not only is the precipitate soluble in an acid solution, but the wool is capable of attracting it from the solution. The precipitate is added to the dye-bath, along with just sufficient oxalic acid to dissolve it, and the wool is dyed in the solution at 100° C., for one and a half hour. Good results are, however, not so readily obtained as when iron and copper mordants are used.

Indigo Substitute.—This product, at present sold in the form of a purplish-blue liquid, is produced by boiling together Logwood extract and chromium acetate. Cotton is dyed by simply working it in a hot solution of the mixture.

Of all the blacks derived from Logwood, the chrome black is the one least affected by acids. If tested by spotting with strong sulphuric acid, it becomes a dark olive colour. It also resists the action of scouring and fulling very well. On the other hand, however, chrome blacks are not altogether satisfactory as regards their behaviour on exposure to light. They gradually assume

LOGWOOD COLOURING MATTERS. 37

a greenish hue, although otherwise they are tolerably fast.

The greening of a chrome black is most apparent when Logwood, or Logwood and Old Fustic, have been employed in dyeing. Its bad effect may be counteracted by the addition of a suitable red colouring matter to the dye-bath —e.g. Alizarin—or by dyeing the wool a reddish-brown colour before dyeing with Logwood. This is very conveniently carried out in practice by boiling the wool with 6-8 % of Camwood for an hour, then adding to the exhausted bath the bichromate of potash (generally with the addition of a small percentage of alum and tartar), and mordanting, etc., as already given.

Owing to the comparatively small proportions of bichromate of potash required to produce the fullest blacks, there is evidently a minimum quantity of lake precipitated on the fibre, so that the latter retains very much its pristine elasticity and softness.

Excess of bichromate of potash should always be avoided, since the colour is then more liable to become green or to fade on exposure to light.

Bichromate of Potash.—The following results of experiments on the use of chromium mordants will be of interest.

By mordanting the wool with 3 % of bichromate of potash a full and bright shade is obtained. The use of more than this amount causes the colour to become dull and grey.

The employment of sulphuric acid along with the bichromate of potash is advantageous when the proportion does not exceed one molecule of sulphuric acid to one molecule of bichromate of potash—*i.e.* 1 % of sulphuric acid, 168° Tw., to 3 % of bichromate of potash.

When used in this proportion it gives a brighter and somewhat deeper shade than can be obtained from bichromate alone; but should the above-mentioned amount be exceeded, a dull grey appearance results, which becomes more apparent as the amount of sulphuric acid increases.

If the bichromate of potash be increased along with the sulphuric acid, the injurious effects of "over-chroming" are intensified.

The addition of tartar or tartaric acid to the mordanting bath, along with bichromate of potash, is beneficial, the shades being much more brilliant, though somewhat lighter, than when sulphuric acid is used.

Tartaric acid gives decidedly brighter and more purple shades than tartar. The best results are obtained by using 6 % of tartaric acid or 8 % of tartar to 3 % of bichromate of potash.

Oxalic acid is also beneficial in the mordanting bath, and in this case 4 % of oxalic acid to 3 % of bichromate of potash yields the best results.

On comparing the shades obtained by using these acids in the mordanting bath, it is seen that they are all better than can be obtained by bichromate of potash alone.

The addition of sulphuric acid produces a deep, dead-looking blue-black; tartar or tartaric acid yields a bright bloomy bluish-black; oxalic acid a black which is darker, duller, and slightly greener than can be obtained with tartar or tartaric acid, but not so dark as with sulphuric acid.

Whenever bichromate of potash alone is employed, the mordanted cloth has a dull yellow colour, but if tartar or tartaric acid has been added to the bath, it is a pale bluish-green.

From these results it would appear that the best shade is obtained when the chromium mordant is fixed on the cloth in the state of chromic oxide previous to the application of the Logwood.

The substitution of chrome alum as a mordant in place of bichromate of potash does not give good results, the ultimate colour obtained having an irregular speckled appearance, evidently owing to the unequal deposition of the chromic oxide; besides, a very large proportion of tartar must be used to obtain a full shade.

When the cloth has been mordanted with bichromate of potash alone, or with bichromate of potash and sulphuric acid, the presence of chalk or calcium acetate in the dye-bath is decidedly injurious. The acetate seems to be least hurtful, although, even with this, the addition of more than 2 % gives the colour a greyish appearance. If tartar has been employed along with the bichromate of potash, the presence of calcium acetate is decidedly beneficial, the shade being intensified from a pale blue when

LOGWOOD COLOURING MATTERS. 39

no calcium acetate is used to a deep indigo-blue when 30 % is employed. The best amount to use appears to be 30 %, but even 80 % may be added to the dye-bath without any great detriment, the colour merely losing a little brilliancy and purple tone, and becoming blacker.

Copperas or Ferrous Sulphate Black.—This-black was formerly the one in general use, but since the introduction of the chrome black it has been more or less discontinued. It is often used for low-class carpet yarns, etc.

Two methods may be employed, namely, that of mordanting the wool first and dyeing afterwards, or that in which the wool is first boiled with Logwood and afterwards saddened.

It is usual to add along with the ferrous sulphate a small proportion of copper sulphate, and when the first method is employed, argol, and frequently also alum, is added.

Example of First Method.—Mordant the wool for 1½-2 hours with 4-6 % of ferrous sulphate, 2 % of copper sulphate, 2 % of alum, 8-12 % of argol; take out, squeeze, and let lie overnight. Dye for 1½ hour with 40-50 % of Logwood.

Example of Second Method.—Boil the wool for one hour with a decoction of 40-50 % of Logwood and 5-10 % of Old Fustic; lift, cool the bath, add 4-6 % of ferrous sulphate, and 2 % of copper sulphate, re-enter the wool, raise the temperature to 100° C. in three-quarters of an hour, and boil half an hour. The first method is the more economical.

The amount of tartar or argol used along with the ferrous sulphate in the first method has considerable influence on the beauty of the colour; with too little it is grey and dull; an excess is less hurtful. Experiment shows that the relative proportions should be 1 molecule of ferrous sulphate, 2-3 molecules of cream of tartar. There is no advantage in using more than 6 % of ferrous sulphate. Wool mordanted with ferrous sulphate alone is buff-coloured from deposition of ferric oxide; when tartar is used its colour remains almost unchanged. If the water employed is not calcareous, the addition of 3 % of chalk, or preferably calcium acetate, to the dye-bath increases the intensity of the colour. The use of a lime salt here does not appear to be so effective as with chromium or aluminium mordants. As with the chrome black, so here the addition of a yellow colouring matter to the dye-bath is necessary in order to obtain a dead-black; without such additions a ferrous sulphate black possesses a bluish-violet hue. The addition of relatively small proportion of Madder, Sumach, etc., aids in giving a fuller and faster black. Sumach, or other tannin matter, when used alone, is incapable of giving a black on wool with ferrous sulphate.

When dyeing unspun wool or yarn it is preferable to use a freshly-made decoction of Logwood, or a good commercial Logwood extract, in order to keep the material clean and free from ground dyewood, since this would interfere in the carding.

The ferrous sulphate blacks become red if spotted with strong mineral acid, and are thus readily distinguished from chrome blacks. They bear the action of scouring and milling satisfactorily, and withstand the action of light better than the chrome blacks. Experiment proves that with regard to fastness against light a simple copper sulphate black is the best, so that the use of copper sulphate along with the ferrous sulphate or potassium dichromate is distinctly beneficial. The use of alum is, on the contrary, detrimental in this respect. The copper sulphate will probably also aid in developing a fuller black by reason of its oxidising action upon the hæmatoxylin.

When employed alone copper sulphate gives greenish shades of blue, having a slightly speckled appearance. The best proportions to employ appear to be 5 % of copper sulphate and 5·5 % of tartar. An excess of tartar causes the shade to become much lighter. With these amounts, and varying the quantity of Logwood, a series of shades, ranging from pale blue to black, may be obtained, but the lighter shades have a distinct greenish appearance, when examined overhand, not observable in the darker shades.

The mordanting and dyeing method yields the deepest and most useful shades. The addition of lime salts to the dye-bath is only slightly beneficial.

Bonsor's Black.—This " direct black," originated by P. Watine-Delespierre, of Lille, consists of a black paste, produced by precipitating a decoction of Logwood with a mixture of ferrous and copper sulphate. It is applied in the same way as the direct chrome black already referred to.

Add to the dye-bath 25-30 % of the black paste and about 2-3 % of oxalic acid. The wool is dyed at 100° C. for 1-2 hours.

It is essential that the solution should not be too acid, or it will not yield its full colouring power. The normal colour of the solution is dark-brown; if blue or green in tint, it is a sign of the presence of undissolved precipitate, and a further slight addition of acid must be made.

As the dyeing proceeds the solution necessarily becomes more and more acid, and it is well, before taking out the wool, to add a small quantity of sodium carbonate to neutralise the excess.

If a deeper shade is wanted, one may add along with the black paste some extract or decoction of Logwood. For a jet-black or dead-black some suitable yellow colouring matter may be added in small quantity, *e.g.* Old Fustic extract, etc. Such additions, however, alter the normal colour of the solution, and a little experience in their use is required.

The spent dye liquor should be kept, and may serve again if replenished with further quantities of black paste and oxalic acid.

It is possible to use this black along with other so-called acid-colours for the purpose of obtaining composite colours, as in the dyeing of unions.

Woaded Blacks are obtained by first dyeing the wool in the indigo-vat to a light or medium shade of blue, then washing well, and dyeing as for chrome or ferrous sulphate blacks. If the chrome-black method is selected, it is advisable to make the addition of tartaric acid to the mordanting bath, in order to reduce to a minimum the oxidising action of the bichromate of potash, and the consequent deterioration of the indigo blue.

Logwood Blues for Wool.—These are much employed by dyers, in order to imitate an indigo-vat blue. They are often combined with the latter by first dyeing the wool a comparatively light blue in the indigo-vat, and then intensifying it by one of the methods now to be described.

Logwood blues are best dyed in two baths, the mordants employed varying in composition and in amount according to the particular tint of blue which it is desired to obtain.

One method is as follows: Mordant the wool for 1-1½ hour, at 100° C., with 4 % of aluminium sulphate, 4-5 % of cream of tartar; wash well, and dye in a separate bath for 1-1½ hour, at 100° C., with 15-30 % of Logwood, and 2-3 % of chalk.

By increasing the amount of alum and tartar the shade is made redder. The addition of the chalk, or preferably calcium acetate, to the dye-bath is very beneficial if the water employed is not calcareous, since it tends to make the colour level, and gives richness and considerable intensity to the blue. When calcium acetate is employed, the best result is obtained by using 30 % of the weight of wool. Some dyers imagine that the use of lime salts in dyeing has only a temporary effect, but this is entirely a mistake; indeed, if distilled water and Logwood liquor be employed, the addition of calcium acetate to the dye-bath becomes just as absolute a necessity for the production of a good full colour as it is for alizarin-red. The colour produced with aluminium mordants is not fast to acids or to light. This difference in respect of fastness to light with the different mordants is somewhat remarkable. A somewhat faster colour is obtained by using along with the alum and tartar 0·5-3 % of bichromate of potash; or, better still, one may mordant entirely with 3 % of bichromate of potash and 1 % of sulphuric acid, 168° Tw. (Sp. Gr. 1·84). If it is desired to imitate the purplish tint of a vat indigo blue, one may then add to the dye-bath, along with the Logwood, a small proportion of Gallein, Alizarin, Galloycyanin, etc. Another method of imparting this purplish "bloom" is to add 0·5-1 % of tin crystals ($SnCl_2 \cdot 2H_2O$) at the end of the dyeing operation.

The brightest Logwood blues are obtained by dyeing at a temperature somewhat below the boil (90° C.). Prolonged boiling tends to dull the colour.

Logwood Purples for Wool.—These are now seldom used. They may be obtained by mordanting the wool with 6 % of tin crystals ($SnCl_2 \cdot 2H_2O$), or its equivalent of muriate of tin, with the addition of 9 % of cream of tartar, and dyeing in a separate bath with 30 % of Logwood. The addition of chalk or calcium acetate to the dye-bath in this case is injurious, since it makes the colour greyer and less intense.

Application to Silk.—The black dyeing of silk has increased to such an enormous extent that some, and even very large, establishments are exclusively devoted to it. Judged from the technical standpoint, it must be admitted that this branch has reached a high standard of excellence, although, on the other hand, it is to be regretted that the practice of weighting silk, which, in the case of black, may reach as high as 600 %, has been so much developed. From 100 kg. = 220 lb. of raw silk the dyer can produce 700 kg. = 1,540 lb. of black silk! The primary object is to increase the volume of the silk fibre, which swells up very considerably, losing, of course, its strength proportionally. The other valuable properties of silk are also more or less deteriorated, and the illusory gain of the buyer is that he requires to pay less for one and the same surface of silk material. On the other hand, fashions now have such a short life that the tendency is to desire something which is cheap and can soon be replaced; so that silk, weighted in reason, has its advocates.

The production of black on silk consists in alternating treatments with iron mordant and tannin matters, with or without a Prussian blue basis.

According to Messrs. Gillet & Son, the present methods of dyeing black silk may be classified in six divisions, as follows:—

BLACK ON BOILED-OFF SILK, 5-15 % loss. A. *Black for Hat Plush.*—1. Mordant in cold nitrate-acetate of iron, and wash. 2. Dye in a decoction of Logwood and a sufficiency of Old Fustic extract. As a rule, 1-2 % of copper acetate and 5-10 % of ferrous sulphate are added to this bath. 3. Dye again in a decoction of Logwood and soap. 4. Brighten in a bath containing a little oil. B. *Masson's Black for Hat Trimmings.—These are* exclusively Parisian goods, and of limited use. The silk is boiled-off in soap containing Logwood decoction, by which means it is rendered less liable to felt. It is mordanted in a solution of partially oxidised ferrous sulphate, with addition of a little copper acetate, and afterwards dyed with Logwood and soap.

C. *English Black*.—Formerly in great demand, this black is now of minor importance. 1. Mordant with basic ferric sulphate, and, after allowing the silk to lie for some time, wash well and soap at 85°-90° C. 2. Dye with 50 % of Old Fustic, 10 % of ferrous sulphate, and 2 % of copper acetate. 3. Dye with Logwood and soap. 4. Brighten.

D. *Black for Velvets*.—The same method as for English black is used, but dye a lighter-coloured black. The tone of colour is frequently modified by giving the silk previously a dark ground of aniline violet or blue. Great care is required in order not to strip off the aniline blue.

BLACK ON BOILED-OFF SILK, original weight or weighted 10 %. E. *Lyons Black* (dating from 1860), for expensive articles.—1. Mordant in a cold strong bath of basic ferric sulphate, 50° Tw. (Sp. Gr. 1·25), once only, and wash. 2. Soap at 85°-90° C. 3. Dye blue with 15-20 % of potassium ferrocyanide and an equal weight of hydrochloric acid, 30° Tw. (Sp. Gr. 1·15). Add the hydrochloric acid in two separate portions. 4. Mordant with basic ferric sulphate, and wash. 5. Give a Catechu bath, 50-100 %, at 60°-80° C. 6. Mordant in a cold solution of alum or aluminium sulphate, and wash. The object of using aluminium mordant is to impart ultimately to the silk a violet or blue-black shade. 7. Dye with Logwood and soap. If shade is too violet, a little Old Fustic is added. 8. Brighten.

F. *Mineral Black* (dating from 1840).—This is a light black, not so fine as the last, and is used for "linings." Mordant with basic ferric acetate, and wash; dye Prussian blue; repeat the mordanting with iron. Prepare with Catechu (100 %) at 80° C. Dye with Logwood and soap. Brighten.

BLACK ON BOILED-OFF SILK, weighted 20-100 % (heavy black). G. This black is dyed on organzine and tram for satins, sarcenets, taffetas, etc. 1. Mordant with basic ferric sulphate, then soap. Repeat these operations 1-8 times, according to the amount of weighting necessary. 2. Dye blue; the proportions of potassium ferrocyanide and hydrochloric acid vary according to the amount of ferric oxide fixed on the silk. 3. Give a Catechu bath

LOGWOOD COLOURING MATTERS. 45

(100-150 %), with the addition of 10-15 % stannous chloride, at 60°-80° C. The employment of stannous chloride in weighted black-silk dyeing has been of the greatest importance, since it facilitates the fixing of the Catechu to a surprising degree, through the formation of a tannate of tin. 4. Give a second bath of Catechu (100-200 %). This is fixed on the silk only by the action of the tin mordant present. 5. Mordant with pyrolignite of iron. 6. Dye with Logwood and soap. 7. Brighten. Blue shades of black are obtained by repeating operations 5, 4, 6, in the order given, four times. The only factors which affect the limitation of weighting are the strength, elasticity, and lustre of the silk itself. As a rule, boiled-off organzine is weighted to 60-70 %, and boiled-off tram to 100 %.

HEAVY BLACK, weighted to 400 %. H. This is used for fringes and the fancy articles of Paris and Lyons; also for the tram silk for satin, cheap ribbons, etc. The raw silk is dyed by working it alternately in chestnut extract and pyrolignite of iron. By repeating these operations fifteen times, the silk is weighted to about 400 %. The final processes consist of brightening operations with 10-20 per cent. of olive-oil. In the first chestnut extract bath, tram is soupled by raising the temperature of the bath sufficiently to soften the silk-glue. Different qualities of silk require slightly different treatment. Bengal silk souples easily; Chinese silk less readily than European silk.

FINE BLACK SOUPLES. I. The finest souples are always obtained by using water as soft as possible, like that of the Gier at Saint-Chamond. 1. Mordant with basic ferric sulphate. 2. Give a soda bath at 30°-40° C. Use 50 % of carbonate of soda crystals. 3. Dye blue with potassium ferrocyanide. 4. Souple by working in a bath of gall-nuts, dividivi, or other similar tannin matter. Heat the bath to 90°-95° C., for 1-3 hours, according to the kind of silk. Experience alone enables the workman to judge when the softening or soupling is sufficient. 5. Leave the silk in No. 4 bath until cold, and then add 5-15 % of stannous chloride crystals. 6. Give a soap bath at 30°-35° C., with 60-80 % of soap. 7. Brighten with 5-15 % of oil. A single iron bath gives 40-50 % of weighting (light souple); two baths give 60-70 %; three give 80 %; four give 80-100 %.

BLACK ON RAW SILK. J. Raw silk is seldom dyed. In order to retain the stiffness of the silk, the silk-glue is not softened, the number of operations is as limited as possible, and the various bath are used at a low temperature. The process consists of working the silk in baths of ferric salt, then in decoctions of Logwood and Old Fustic. For boiled-off silk the potassium ferrocyanide baths are employed at a temperature of 55°-60° C., because the formation of Prussian blue would otherwise proceed only very slowly, creeping, as it were, from the periphery to the centre of the fibre. For souple silks, however, it is necessary to use cold baths. Only a portion of the requisite hydrochloric acid is at first added, in order to avoid dissolving off any basic ferric sulphate.

The operation of brightening is intended to restore the soft feel and lustre, which have been greatly destroyed by reason of the large amount of foreign matter with which the silk has become encrusted. For boiled-off silk there is used about 1-2 % of olive oil; for souples, 5-15 %; and for fringes, etc., 5-20 %. The oil is made into an emulsion with carbonate of soda at 60°-70° C., or with caustic potash or soda in the cold, and then immediately mixed in the bath with water. The silk must be worked in the mixture at once, *i.e.* before any separation of the oil can take place. Very often an addition is made of 40-60 % of citric, tartaric, or acetic acid.

The usual duration of each operation varies from 1-2 hours, but in the tannin baths the silk should remain longer; it is frequently left steeping in them over-night.

The black dyeing of Tussur silk is a difficulty not yet completely overcome. The shades are not satisfactory, and the fibre becomes covered with metallic-looking spots.

1. Boil-off with dilute caustic soda. 2. Mordant once or twice in basic ferric sulphate, and fix by means of a bath of weak caustic soda. 3. Dye with potassium ferrocyanide. 4. Prepare with a weak chestnut extract bath. 5. Mordant with pyrolignite of iron, and repeat operations 4 and 5. 6. Brighten with 6-8 % olive oil.

CHAPTER III.

NATURAL RED AND YELLOW COLOURING MATTERS.

Brazilwood, Peachwood, and Limawood.—Obtained from various species of *Cæsalpinia*, and their dyeing properties were similar, but owing to the fugitive colours they gave they were seldom used except for shading purposes, and have now been largely replaced by artificial dyes. Application to cotton : With aluminium mordants dull bluish-red colours were obtained; stannic mordants gave brighter and more orange-toned reds; and iron mordants gave violet-grey shades. Application to wool A mordant of bichromate of potash gave a purplish-slate colour if used in small quantities, and a claret brown with larger quantities; aluminium sulphate gave a bluish-red colour, which could be made bluer by ammonia; stannous and stannic chloride yielded bright reds; copper sulphate, drab or claret brown shades; and ferrous sulphate dark slate and claret. Application to silk : Crimson shades were obtained with alum, and bright crimsons by stannous chloride.

Camwood, Barwood, and Sanderswood.—Obtained from certain species of *Pterocarpus* and *Baphia*, and are similar in dyeing properties but different in tone. Application to cotton : Aluminium and tin mordants gave good reds, and good chocolate or violet brown shades were obtained with pyrolignite of iron. Application to wool Rich claret brown shades were obtained with bichromate of potash, dull brownish-red shades with aluminium, brighter and bluer reds with stannous chloride, claret brown with copper sulphate, and purplish shades with ferrous sulphate.

Madder.—Obtained from the dried roots of *Rubia tinctorum*, and although once perhaps the most useful dyestuff, has now been entirely replaced by artificial Alizarin.

Cochineal.—Obtained from the dried insect *Coccus*

cacti, but has been replaced by the azo-reds, and is now only used in exceptional cases.

Orchil.—Prepared from certain lichens by oxidising them in the presence of ammonia. It is still sometimes used for wool and silk, being dyed with the former with a little sulphuric acid, and the latter in a soap solution. The colour produced is a bright bluish-red or magenta.

Annatto.—Obtained from the pulp surrounding the seeds of *Bixa orellana*, and is only used to a very limited extent in silk dyeing with the addition of soap. Pale shades are obtained at 50° C. and dark ones at 80°-100° C., but they are of a fugitive character.

Safflower.—The dried florets of *Carthamus tinctorius*, and yields a bright but fugitive pink on all fibres.

Weld.—The plant *Reseda luteola*, and was used for yellow and olive colours on wool and silk, but chiefly for silk, and for certain colours it is still used on that fibre. It is dyed on silk which has been mordanted with alum with 20-40 % Weld at 50°-60° C.

Old Fustic.—The wood of *Morus tinctoria*, and was never used much for cotton. Applied to wool: With bichromate of potash old gold shades are obtained; aluminium mordant gives yellow shades, and bright fast colours can be obtained with stannous mordant. It is applied to silk for a light yellow by working for $\frac{1}{4}$-$\frac{1}{2}$ hour, at 50°-60° C., in a bath containing 16 % alum and 8-16 % Old Fustic.

Quercitron Bark.—The inner bark of *Quercus tinctoria*, and was never used much for cotton or silk dyeing. Applied to wool Bichromate of potash gives reddish-olive yellows, aluminium yellows are paler, and stannic chloride gives a pale buff colour.

Flavin.—A preparation of Quercitron Bark which gives stronger colours.

Young Fustic.—The wood of the sumach tree (*Rhus cotinus*) which was once used largely in silk dyeing, and still is very occasionally used for certain oranges and scarlets on wool. Bichromate of potash gives a reddish brown, aluminium mordants produce yellowish buffs, bright yellows are obtained with stannous mordants, whilst copper and iron mordants give olive shades.

Persian Berries.—The dried unripe fruit of *Rhamnus*,

YELLOW COLOURING MATTERS.

which were chiefly used for printing olive, orange, and green shades on cotton.

Turmeric.—The tuber of *Curcuma tinctoria*, which yields a bright yellow on cotton without the aid of a mordant. It can also be used on wool without a mordant, but gives darker shades with aluminium and more orange ones with tin.

Barberry.—The root and bark of *Berberis vulgaris*, and was never used beyond silk and leather dyeing, giving light shades with sulphuric, acetic, or tartaric acids, and dark ones with stannous chloride.

Catechu.—Obtained from species of acacia, areca, and uncaria, and was once largely used for various brown to black shades on cotton and for black on silk. It has not yet been entirely replaced by artificial dyestuffs. On cotton it is applied with bichromate of potash or aluminium, but it was never much used for wool.

CHAPTER IV.

ANILINE COLOURING MATTERS.

RECENT years have brought about a great change in dyeing methods, this being caused not only by the introduction of artificial colouring matters but by the way in which it has been possible to substitute their use in practically every case where natural colouring matters were previously employed. One by one the old dyes have been supplanted, and it is only in certain cases where their use has its advantages for obtaining some special result. Even natural Indigo, which only recently was supposed to be unique for certain purposes and quite impossible to replace, is now finding its place gradually usurped by synthetic Indigo. There are a large number of artificial dyestuffs which are cheap but fugitive, and these find their uses in low goods which are not expected to be long in use, but, on the other hand, the colours previously obtained by natural dyestuffs can now be obtained, in many instances, by cheaper and easier methods and with better results.

Although there are now hundreds of brands of artificial dyes, these are being added to at the rate of a few a week, so that no attempt will be made here to accomplish the impossible task of enumerating them all. The earlier and principal brands are therefore mentioned, as these not only form the basis of other and later colours, but many so-called new dyes are nothing but mixtures of old ones.

(a.) Rosaniline Group.

Benzaldehyde Green. $\quad C \begin{cases} C_6H_4 \cdot N(CH_3)_2 \\ C_6H_5 \\ C_6H_4 \\ N(CH_3)_2Cl \end{cases}$

This colouring matter is derived from dimethyl-aniline, and this is the hydrochloride of tetra-methyl-diamido-triphenyl-carbinol. In commerce it occurs really as various salts (sulphates, oxalates, etc.), or zinc double salts of the colour-base, and is sold under a variety of names,

ANILINE COLOURING MATTERS.   51

e.g. Malachite Green [$3(C_{23}H_{24}N_2\cdot HCl)2ZnCl_2\cdot 2H_2O$] (Berlin Aniline Co.), Solid or Fast Green [$2C_{23}H_{24}N_2\cdot 3C_2H_2O_4$] (L. Cassella & Co.), Victoria Green (Badische Anilin and Soda Fabrik), Benzoyl Green, New Green, etc. Closely allied colouring matters possessing similar dyeing properties are those derived from diethyl-aniline, and known as Ethyl Green, Fast Green J, New Victoria Green OG (BASF), Brilliant Green, etc. They give yellower shades of green than those derived from dimethyl-aniline. In dyeing with all these greens, the temperature of the dye-bath may, if necessary, be raised to 100° C. without injuring the colour.

Application to Cotton.—Prepare the cotton with tannic acid and tartar emetic, wash, and dye in a fresh bath. The dye-solution may be cold or it may be heated gradually to a temperature not exceeding 60° C. Add the colour solution (2-4 %) gradually. The dye-bath, which is sometimes very slightly acidified with acetic acid, is not exhausted, and should be preserved for fresh lots of material.

For bright shades the cotton may be prepared with sulphated oil and aluminium sulphate.

For yellow shades of green add to the dye-bath Auramine or other basic yellow colouring matter. One may also mordant with aluminium, fix with phosphate of soda, and dye first with Quercitron Bark and then, in a separate bath, with Benzaldehyde Green.

Jute is dyed with this and all basic colouring matters without being mordanted.

Application to Wool.—Dye in a neutral bath, that is, without any addition other than the colour-solution. The custom of adding soap to the dye-bath is not to be recommended, because of the formation of insoluble sticky zinc soap. An addition of at most 1-2 % of sulphuric acid, 168° Tw. (Sp. Gr. 1·84), alum, or other acid salt, may be made under special circumstances. The colour shows a great tendency to rub off, and does not withstand the action of milling or of light. Better results in all these respects are obtained by mordanting the wool previously with thiosulphate of soda, according to the method given for Methyl Green.

Application to Silk.—Dye at a temperature of 50°-60° C. in a bath containing a small quantity of soap, or "boiled-off" liquor. Sometimes a little sulphuric acid is added to the bath. Wash well, brighten in a cold bath very slightly acidulated with acetic acid, and dry.

For yellow shades of green, add to the dye-bath Auramine or other basic yellow colouring matter. If Picric Acid or other acid colour is used, it must be added to the brightening bath, taking care not to use any excess of acetic acid, otherwise the green will be stripped off. It is better to dry the silk at once after brightening, in order not to remove any Picric Acid by rinsing.

Acid Green.—Several acid greens are met with in commerce, bearing such names as Helvetia Green (Soc. Chem. Ind., Basle), Acid Green (Poirrier), Light Green S (BASF), Guinea Green (Berlin Aniline Co.), etc. They are sodium or calcium salts of the sulphonic acids of the base of one or other of the Benzaldehyde Greens.

As their name implies, they require to be applied in an acid bath. Their dyeing power is less than that of the Benzaldehyde Greens, but, employed with other acid colours, they are extremely useful for producing compound shades. They are not suitable for dyeing cotton.

Application to Wool.—Dye with the necessary amount of colour and 2-3 times its weight of sulphuric acid, 168° Tw. (Sp. Gr. 1·84), added. An addition of 10-20 % of sodium sulphate may also be made if there is any tendency to uneven dyeing. Raise the temperature gradually to near the boiling point.

Application to Silk.—Dye at a temperature of 50° C. with the addition of "boiled-off" liquor slightly acidified with sulphuric acid.

Alkali Green, Viridin.—This colouring matter, derived from diphenylamine by the Malachite Green process or prepared by oxidising benzyl-diphenylamine with chloranil, is only of limited use. Being a sodium sulphonate of the colour-base, its dyeing properties are similar to those of Alkali Blue, and it is applied in exactly the same way. Pleasing greenish shades of blue may be obtained on wool by using mixtures of Alkali Blue and Alkali Green.

$$\textit{Magenta.} \quad C \begin{cases} C_6H_4 \cdot NH_2 \\ C_6H_4 \cdot NH_2 \\ C_6H_3 \cdot CH_3 \\ NH_2 \cdot Cl \end{cases}$$

ANILINE COLOURING MATTERS.

This colouring matter, sometimes called Roseïne, Ponceau, Rubine, etc., is usually the hydrochloride or the acetate of the organic base tolyl-diphenyl-triamido-carbinol or rosaniline. Various bye-products, containing impure Magenta, are sold under different names, *e.g.* Cerise, Grenadine, Cardinal, Amaranth, etc. None of these yield dyes which are fast to light.

Application to Cotton.—Prepare the cotton with tannic acid, and either tartar emetic or stannic chloride, wash well, and dye in a separate bath at 45°-50° C. Add the colour solution to the bath gradually. In conjunction with other basic colouring matters—*e.g.* Chryoïdine, Neutral Violet, etc.—numerous compound shades (claret, bordeaux, etc.) can be obtained.

Brighter colours are obtained by mordanting the cotton with sulphated oil and aluminium sulphate, but they are not so fast to soap as those fixed by means of tannic acid. Both methods are applicable to all basic colouring matters.

Application to Wool.—Dye in a neutral bath. Heat gradually to 90° C. With the addition of 2-4 % of soap, the colour obtained is brighter; but, in this case, the bath cannot be exhausted, and should be preserved, if possible, for subsequent lots of woollen material. This is the general mode of dyeing with all basic colouring matters.

The colour bleeds in milling, and is not suitable for "Tweed-yarns," etc.

Application to Silk.—Dye at 50°-60° C. in a weak, fresh, soap bath, adding the colour solution gradually. Wash and brighten, for yellow shades, in a bath slightly acidulated with acetic or tartaric acid; for blue shades acidify with sulphuric acid.

Acid Magenta.—This colouring matter is the sodium salt of the trisulphonic acid of rosaniline. It has only about half the colouring power of Magenta, and is not applicable in cotton-dyeing.

Application to Wool.—Dye in a bath acidified with 2-4 % of sulphuric acid, 168° Tw. (Sp. Gr. 1·84), with the addition of 20-30 % of sodium sulphate if there is any tendency to irregular dyeing. Introduce the textile material at 40° C., raise the temperature to 100° C. in 40 minutes, and boil 20-30 minutes. This is the general mode of dyeing with all similar acid colours.

Acid Magenta is much used in conjunction with other colouring matters applied in an acid bath for dyeing compound shades. The colour it yields is decolorised by the action of alkalis, and is not suitable for goods requiring to be milled.

Application to Silk.—Dye in a bath containing "boiled-off" liquor slightly acidulated with sulphuric acid. Add the colour solution gradually.

Spirit Blues: Rosaniline Blue. $\quad C \begin{cases} C_6H_4 \cdot NH(C_6H_5) \\ C_6H_4 \cdot NH(C_6H_5) \\ C_6H_3 \cdot CH_3 \\ NH \cdot C_6H_5 \cdot Cl \end{cases}$

This colouring matter, in its purest form, is the hydrochloride of tri-phenyl-rosaniline. According to its purity, this blue yields shades which vary from a dull reddish-blue to a pure sky-blue. The redder shades are marked R (Dahlia, Parma Blue, etc.), while those marked 5 B and 6 B (Opal Blue, Lyons Blue, etc.), give the purer shades. Intermediate products are marked B, 2 B, 3 B, 4 B (Humboldt Blue, Imperial Blue, Gentiana Blue, etc.).

Diphenylamine Blue. $\quad C \begin{cases} C_6H_4 \cdot NH(C_6H_5) \\ C_6H_4 \cdot NH(C_6H_5) \\ C_6H_4 \\ NH \cdot C_6H_5 \cdot Cl \end{cases}$

This colouring matter, derived from diphenylamine, is considered as the hydrochloride of tri-phenyl-para-rosaniline. It is a somewhat finer but also more expensive blue.

Methyl and Ethyl Blue.—These are methyl and ethyl derivatives of Diphenylamine Blue, and are distinguished by the extreme purity of the greenish-blue colour which they yield.

These spirit colours are dissolved in 40-50 times their weight of methylated spirit, sometimes with the addition of a very little sulphuric acid. Use a narrow-necked vessel, and heat by placing it in hot water.

Application to Cotton.—The cotton is prepared with oleate of alumina. Work the bleached cotton in a hot (60° C.) soap bath (60-100 g. = 1·9-3·2 ozs. of soap per kg. = 2·2 lb. of cotton), squeeze, and work in a cold bath of aluminium acetate, 8° Tw. (Sp. Gr. 1·04), and squeeze.

ANILINE COLOURING MATTERS. 55

Repeat these operations three times, and dye in a fresh bath. One may also work the soaped cotton at once in a dye-bath to which aluminium acetate has been added. Add the colour solution in small portions, and heat gradually to the boiling point. Wash well in cold water, and pass finally through a weak soap bath, heated to 60° C., to which a little acetic acid has been added till it just begins to show signs of turbidity.

Cotton may also be prepared with tannic acid, and dyed in a fresh bath containing colour-solution and acidified with alum.

Application to Wool.—Dye with 1-5 % of colouring matter (or more if necessary), with the addition of 4-8 % of sulphuric acid, 168° Tw. (Sp. Gr. 1·84), and 10-20 % of sodium sulphate. Enter the wool at 50°-60° C., heat up rapidly to 100° C., and boil ½ hour, or longer if necessary. The acid and also the colour solution should be added gradually to the bath, and in small portions at a time, in order to ensure a regular colour. Instead of sulphuric acid one may also acidify with 10-12 % of alum.

When red shades of blue are used, the addition of sulphuric acid gives the best colour; with the purer blues brighter shades are obtained with the use of alum.

Owing to their insolubility, Spirit Blues are very apt to dye unevenly, but they are preferred by yarn-dyers when the goods have subsequently to be milled.

Application to Silk.—Introduce the silk into a tepid bath containing "boiled-off" liquor acidified with sulphuric acid. Add the colour solution by degrees, heat gradually to 100° C., and dye at this temperature. Wash in cold water and brighten with dilute sulphuric acid.

Soluble Blues.—The Soluble Blues generally consist of the ammonium or sodium salts of the *di*- and *tri*-sulphonic acids of Rosaniline or Diphenylamine Blues. They vary considerably in purity of colour, and are marked R, B, 2 B, 6 B, etc. The redder shades are known by such commercial names as Serge Blue, Navy Blue, Blackley Blue, etc., while those of purer tone are known as China Blue, Night Blue, Soluble Blue, Water Blue, Cotton Blue, etc.

Application to Cotton.—The cotton is prepared with tannic acid and tartar emetic, and dyed at 60°-70° C., in a separate bath, slightly acidulated with alum.

Light shades of blue are frequently dyed without any previous preparation.

The method given for Spirit Blue may also be used.

Another method recommended is to work the cotton at 60° C. in a bath containing the colour solution and 3 % of stannate of soda. When the cotton is properly saturated, acidify the solution with sulphuric acid, and work ½ hour longer.

In all cases the dye-bath should be preserved, since not only is the bath never completely exhausted, but further lots dyed in the same bath have also a brighter colour.

Application to Wool.—Dye as in the same manner as with Spirit Blues.

Application to Silk.—Dye as with Spirit Blues.

Alkali Blue.—This is a special kind of Soluble Blue; it consists essentially of the sodium salt of the *mono-*sulphonic acid of Rosaniline Blue. Both red and blue shades are met with; they are marked accordingly with R, B, etc., or they have such names as Guernsey Blue, Fast Blue, Nicholson's Blue, etc. The red shades are faster than the blue shades, both to light and to the action of dilute alkalis in milling, etc.

Alkali Blues are not applicable in cotton-dyeing.

Application to Wool.—Wool must always be dyed with Alkali Blue in a slightly alkaline bath (hence the name). The wool takes it up in the form of a colourless sodium salt. The development of the colour—*i.e.* the precipitation on the fibre of the blue-coloured mono-sulphonic acid—is effected in a separate and slightly acid bath. The water of the dye-bath should be as free from lime-salts as possible, since the lime compound of the mono-sulphonic acid is insoluble.

In dyeing wool, add to the dye-bath the amount of colour solution requisite to obtain the desired shade (say 0·5-5 %), and dissolve in it also 4-8 % of carbonate of soda (crystals). Introduce the wool into the bath at 40° C., heat rapidly to 80° or 100° C., and boil ½-¾ hour. It is then taken out, washed well in water, and transferred to a bath containing water slightly acidulated with sulphuric acid (5 % of sulphuric acid, 168° Tw.). It is worked in this bath for 15-20 minutes at 60° C., until the colour is fully developed, and is then washed free from acid. If the first bath, or the dye-bath, as it may be called, is sufficiently alkaline, the wool acquires therein only a very pale bluish tint, but on passing it into the second bath, which may be named the acid or developing bath, the blue is at once developed. The dye-bath is never exhausted, and should always be preserved. For succeeding lots of wool add proportionately less of both colour solution and carbonate of soda. Instead of carbonate of soda, borax or strong ammonia solution may be used. Under no circumstances whatever should the sulphuric acid be added to the dye-bath, otherwise the colour-acid will be precipitated. In order to preserve uniformity of dye in several lots of material, it is very advisable not to develop different shades of blue in the same acid-bath. The acid-bath should never be employed at a temperature above 80° C., otherwise the blue is less brilliant.

In order to match-off any given shade, a small sample of the woollen material must occasionally be taken from the bulk during the dyeing operation, and passed through warm dilute acid to develop the blue. Only in this way is it possible to regulate what amount of colour solution should be added to the dye-bath, and to determine the duration of the dyeing operation.

Many Alkali Blues are not improved in colouring power by an addition of carbonate of soda to the dye-bath; they seem to be already sufficiently alkaline, while others give even inferior colours if the addition is made. In all cases a large excess of alkaline salt is to be avoided, since it tends to impoverish the colour.

Colours said to be somewhat faster to milling are obtained by adding zinc sulphate or alum to the acid-bath, since these salts form insoluble lakes with the colouring matter.

Alkali Blue may be employed for the purpose of obtaining compound colours, by using it in conjunction with various acid-colours—*e.g.* Croceïn Scarlet or Orange, etc. In these cases the acid bath is dispensed with, and after dyeing with Alkali Blue and washing, the wool is at once worked in the dye-bath of the acid-colour, with the necessary additions. The development of the blue, and the dyeing of the scarlet, etc., are thus simultaneous.

Application to Silk.—Silk is dyed like wool, but it is preferable to use borax in the dye-bath instead of carbonate of soda or ammonia.

$$\text{Rosaniline Violets.} \quad \text{C} \begin{cases} C_6H_4 \cdot NH(C_6H_5) \\ C_6H_4 \cdot NH(C_6H_5) \\ C_6H_3 \cdot CH_3 \\ NH_2 \cdot Cl \end{cases}$$

These violets, also called by such names as Phenyl Violet, Spirit Violet, Parma Violet, Imperial Violet, etc., are hydrochlorides of mono- and di-phenyl-rosaniline. They find now only a limited use, being less bright than the Methyl Violets; they are, however, said to be somewhat faster to light and to milling, and may be used with advantage when a dull, moderately-fast violet is required, as in felt-hat dyeing.

Closely related to Rosaniline Violet is the so-called Regina Purple (Brook, Simpson, and Spiller).

Application to Cotton.—Prepare the cotton with tannic acid and tartar emetic, and dye in a bath slightly acidulated with sulphuric acid or alum.

Application to Wool.—Dye at 60°-80° C. in a colour solution acidulated with 4 % of sulphuric acid, 168° Tw. (Sp. Gr. 1·84). Since these are basic colouring matters, the need of acidulating the bath is noteworthy.

Application to Silk.—Dye at 60°-80° C. in a bath containing boiled-off liquor, slightly acidulated with sulphuric acid.

$$\text{Hofmann's Violet.} \quad \text{C} \begin{cases} C_6H_4 \cdot NH \cdot CH_3 \\ C_6H_4 \cdot NH \cdot CH_3 \\ C_6H_3 \, CH_3 \\ NH \cdot CH_3Cl \end{cases}$$

This colouring matter, also called Dahlia, Primula, etc., is considered as the hydrochloride of the base trimethyl-rosaniline; it is only used for red shades of violet, the bluish violets being better obtained from the Methyl Violets. The colour it yields is not fast to light.

Application to Cotton.—Prepare the cotton with tannic acid and tartar emetic, or with sulphated oil and aluminium acetate; wash and dye at 45°-50° C. in a neutral bath.

Application to Wool.—Dye at 60°-80° C. in a neutral bath, or with the addition of 2-4 % of soap.

Application to Silk.—Dye at 50°-60° C. in a bath containing soap or "boiled-off" liquor, with or without the addition of a little sulphuric acid. Wash and brighten in a bath slightly acidulated with acetic or tartaric acid.

$$\text{Methyl Violet.} \quad \text{C} \begin{cases} C_6H_4 \cdot N(CH_3)_2 \\ C_6H_4 \cdot N(CH_3)_2 \\ C_6H_4 \\ NH \cdot CH_3 \cdot Cl \end{cases}$$

This colouring matter, also called Paris Violet, is considered as the hydrochloride of penta-methyl-para-rosaniline. Various brands are sold—as Methyl Violet R, B, 3 B, etc.—according as they yield red or blue shades of violet.

Some of the Methyl Violets are zinc double salts, and are then sold in the crystalline state; with these the addition of soap to the dye-bath must be strictly avoided.

The methods of applying them to the textile fibres are identical with those employed for Hofmann's Violet.

Benzylrosaniline Violet.—This colouring matter is the benzyl (C_7H_7) derivative of Methyl Violet. The most highly benzylated product is generally sold as Methyl Violet 6 B, and by mixing this in different proportions with Methyl Violet B, the various marks of Methyl Violet 2 B, 3 B, 4 B, 5 B are obtained.

Benzyl Violet yields much bluer shades of violet than Methyl Violet, although the method of its application to the various fibres is very similar. It bears the addition of a little sulphuric acid to the dye-bath better than Methyl Violet.

Alkali Violet (Meister, Lucius & Brüning) is applied in the same manner as Alkali Blue.

$$\begin{array}{c} \textit{Acid Violet 4 RS.} \\ (\text{BASF}). \end{array} \quad \text{C} \begin{cases} C_6H_2(CH_3) \cdot NH(CH_3)SO_3Na \\ C_6H_3 \cdot NH(CH_3)SO_3Na \\ C_6H_3 SO_3Na \\ NH \end{cases}$$

This colouring matter is the sodium salt of di-methylrosaniline-tri-sulphonic acid.

Acid Violet 5 RS (BASF) is the corresponding monomethyl compound.

Acid Violet 6 B (BASF) is the corresponding benzylmethyl compound.

These colouring matters, sold by several manufacturers with different brands, are adapted only for wool and silk, and are applied in the same manner as Acid Magenta. They are useful, in conjunction with other acid colouring matters, for producing compound shades.

Methyl Green. $\quad \mathrm{C} \begin{cases} \mathrm{C_6H_4 \cdot N(CH_3)_2} \\ \mathrm{C_6H_4 \cdot N(CH_3)_2 \cdot CH_3 \cdot Cl} \\ \mathrm{C_6H_4} \\ \mathrm{NH \cdot CH_3 \cdot Cl} \end{cases}$

This colouring matter may be regarded as the methyl chloride compound of Methyl Violet. It occurs in commerce in the crystalline state as a zinc double salt. Although still used, it has been very largely supplanted by the Acid and Benzaldehyde Greens, since these are much cheaper, and offer certain advantages in point of application and stability.

Application to Cotton.—Dye in the same manner as with Benzaldehyde Green.

Application to Wool.—Owing to the weak attraction which wool has for Methyl Green, it is necessary that it should be mordanted previous to being dyed with this colouring matter. The ordinary mordants, however, are of no use, and recourse is had to the singular and strong affinity which amorphous sulphur has for Methyl Green.

The wool is mordanted in a bath containing 10-20 % of thiosulphate of soda (usually called hyposulphite of soda) and acidified with 5-10 % of sulphuric acid, 168° Tw. (Sp. Gr. 1·84), or hydrochloric acid, 32° Tw. (Sp. Gr. 1·16). Introduce the wool into the milky liquid at 40° C., raise the temperature gradually to 80° C. in the course of 1 hour, then wash well. Dye at 50°-60° C. in a separate bath containing Methyl Green and 2-4 % of borax or acetate of soda.

The addition of these latter salts to the dye-bath has the effect of neutralising the acid remaining in the wool after washing; if, however, previous to dyeing, the wool is worked for about ¼ hour, at 70° C., in a weak solution of carbonate of soda or ammonia, their addition to the dye-bath is unnecessary.

The shade produced by Methyl Green is always bluish, and if the temperature of the dye-bath is raised to 100° C. it becomes still bluer, owing to a portion of the colouring matter decomposing at this temperature with elimination of methyl chloride and the production of Methyl Violet;

the effect obtained is that of a mixture of green and violet, namely, blue (Peacock Blue). If it is desired to obtain yellower shades of green, Picric Acid may be added to the dye-bath, but since this only dyes in an acid-bath (a condition which is prejudicial to the dyeing property of Methyl Green) one must add also a small proportion of acetate of zinc. This salt is gradually decomposed by the sulphur already fixed on the wool, and the liberated acetic acid causes the Picric Acid to dye, while it does not prevent the Methyl Green from doing so. The zinc sulphide produced acts as a mordant for the Methyl Green in the same manner as the sulphur. Should, however, the Methyl Green dye slowly, from over-acidity of the bath, the addition of a little acetate of soda is necessary.

It is essential that in the operations of mordanting and dyeing the use of metal, either in the dye-vessels themselves or in the utensils employed, should be strictly avoided, otherwise the wool may acquire a dark colour, or be spotted, by the production of metallic sulphides.

Application to Silk.—Dye exactly as in the case of Benzaldehyde Green.

Auramine (BASF) (Soc. of Chem. Ind., Basle). $[C_6H_4N(CH_3)_2]_2 \cdot C \cdot NH \cdot HCl$.—This yellow colouring matter is the hydrochloride of tetra-methyl-diamido-benzo-phenonimide. It is particularly useful to the cotton-dyer, and is said to resist the action of light and soap solutions fairly well, but is readily affected by chlorine. It should be dissolved in hot water, but the solution should not be boiled, since the colouring matter is thereby decomposed. It is useful for producing compound shades in conjunction with other basic colouring matters, as Safranine, Benzaldehyde Green, etc.

Application to Cotton.—Mordant the cotton with tannic acid and tartar emetic, and dye in a separate bath. Introduce the cotton into the cold colour solution, and raise the temperature of the bath to 40°-50° C.

Application to Wool.—Dye in a neutral bath. Enter cold and heat gradually to 70° C. Better colours are said to be obtained if the wool is previously mordanted with sulphur, after the manner in vogue for Methyl Green.

Application to Silk.—Dye in the same manner as with Magenta.

Ethyl Purple 6 B (BASF). $\text{C}\begin{cases} \text{C}_6\text{H}_4\cdot\text{N}(\text{C}_2\text{H}_5)_2 \\ \text{C}_6\text{H}_4\cdot\text{N}(\text{C}_2\text{H}_5)_2 \\ \text{C}_6\text{H}_4 \\ \text{N}(\text{CH}_3)_2\text{Cl} \end{cases}$

This colouring matter is the hydrochloride of hexa-ethyl-para-rosaniline. It is applied to the various fibres in the same way as Hofmann's Violet.

Crystal Violet 5 BO (BASF). $\text{C}\begin{cases} \text{C}_6\text{H}_4\cdot\text{N}(\text{CH}_3)_2 \\ \text{C}_6\text{H}_4\cdot\text{N}(\text{CH}_3)_2 \\ \text{C}_6\text{H}_4 \\ \text{N}(\text{CH}_3)_2\text{Cl} \end{cases}$

This colouring matter is the hydrochloride of hexa-methyl-para-rosaniline. It is applied to various fibres like Hofmann's Violet, over which it possesses the advantage of greater colouring power, of extreme solubility in water, and of having no tendency to produce a bronze scum on the surface of the dye liquor or on the dyed material.

Victoria Blue 4 R (BASF). $\text{C}\begin{cases} \text{C}_6\text{H}_4\cdot\text{N}(\text{CH}_3)_2 \\ \text{C}_{10}\text{H}_6\cdot\text{N}(\text{CH}_3)(\text{C}_6\text{H}_5) \\ \text{C}_6\text{H}_4 \\ \text{N}(\text{CH}_3)_2\text{Cl} \end{cases}$

This colouring matter is the hydrochloride of penta-methyl-phenyl-triamido-diphenyl-α-naphthyl-carbinol. It is applied to the various fibres in the same manner as Hofmann's Violet. Wool and silk may be dyed with the addition of a little acetic or sulphuric acid to the bath, in the same manner as acid colours. The dyeing power is thereby somewhat lessened and the bath is not so well exhausted, but the colour obtained seems brighter.

Victoria Blue B (BASF). $\text{C}\begin{cases} \text{C}_6\text{H}_4\cdot\text{N}(\text{CH}_3)_2 \\ \text{C}_{10}\text{H}_6\cdot\text{NH}\cdot(\text{C}_6\text{H}_5) \\ \text{C}_6\text{H}_4 \\ \text{N}(\text{CH}_3)_2\text{Cl} \end{cases}$

This is the tetra-methyl compound corresponding to Victoria Blue 4 R, and may be applied in the same way.

Night Blue (BASF).—This colouring matter is closely related to the last, and is applied to the textile fibres in a similar manner. It requires to be dissolved in dilute acetic acid to prevent decomposition on boiling.

Phosphine [$\text{C}_{20}\text{H}_{17}\text{N}_3\cdot\text{HCl}$].—This orange colouring matter (said to be quinoline derivative) is the hydro-

ANILINE COLOURING MATTERS. 63

chloride of the base chrysaniline. Its dyeing properties are similar to those of Magenta, and it is applied to the textile fibres in the same manner.

It finds only a limited use in wool- and silk-dyeing, because of its expense.

Rosolane [$C_{27}H_{24}N_9 \cdot HCl$].—This colouring matter is the hydrochloride of the base mauveïne; it is, indeed, the original Perkin's Violet.

Its method of application is similar to that of the Methyl Violets. Although itself not requiring a mordant, it may be used in conjunction with polygenetic colouring matters for the production of compound shades. It is used as a substitute for Orchil or Ammoniacal Cochineal in the production of bright greys.

(b.) *Induline and Safranine Group.*

Indulines.—These comprise a number of colouring matters made by different processes, but all possessing somewhat similar dyeing properties. They are known by a variety of commercial names, *e.g.* Violaniline, Nigrosine, Elberfeld Blue, Bengaline, Aniline Grey, Coupier's Blue, Roubaix Blue, etc.

Those used for cotton-dyeing are insoluble in water, and require to be dissolved in methylated spirit. These Spirit Indulines are hydrochlorides of a colour-base, as violaniline, triphenyl-violaniline, etc. For wool- and silk-dyeing they are treated with strong sulphuric acid; they are thus rendered soluble in water, and are sold as sodium salts of the corresponding sulphonic acids.

They all yield dark, dull blue colours, not unlike indigo-vat blues, to imitate which they are frequently employed.

Application to Cotton.—Prepare the cotton with tannic acid and tartar emetic, wash, and dye in a separate bath containing the colour solution, acidified slightly with sulphuric acid or by the addition of alum (10 %). Dye at a temperature of about 60° C. The bath is not exhausted, and must be preserved for succeeding lots of material. One may also employ the indigo-vat method.

Application to Wool.—Owing to the precipitation of the free sulphonic acids of these colours on the addition of acid to their solutions, it is extremely difficult to dye light shades evenly with them. They are best adapted for dyeing dark shades.

Add the requisite amount of colour-solution (5-15 %) to the dye-bath, heat to 100° C. as rapidly as possible, enter the wool, and boil 1-1½ hour, without any other addition. Continue now to boil 1 hour longer, during which period add from time to time dilute sulphuric acid in small portions. Use 5-15 % of sulphuric acid, 168° Tw. (Sp. Gr. 1·84), according to the amount of colouring matter employed.

The long boiling with colour solution alone enables the wool to become thoroughly permeated with the colouring matter while still in the soluble state. An addition of 5-10 % of borax, carbonate of soda, or strong ammonia solution at this stage is beneficial. The actual dyeing of the wool begins only when the bath is acidulated; the addition of acid should always be made slowly, so that the wool may take up the gradually precipitated colouring matter as evenly as possible.

Wool is said to dye much better with Induline if it has been previously rinsed in a weak solution of bleaching-powder and then in dilute hydrochloric acid.

These colours have been frequently recommended as good substitutes for indigo-vat blues. Although fairly fast to light, they gradually lose their bluish tint and brilliancy on exposure, and assume a dull greyish tone. Towards weak alkalis they are moderately fast; the action of acids they withstand perfectly. In conjunction with other acid-colours they are useful for producing a large variety of compound shades.

Application to Silk.—Dye in a bath containing " boiled-off " liquor, acidified slightly with sulphuric acid. Enter the silk at 60° C., add the colour solution gradually, raise the temperature gradually to 100° C., and boil ½ hour. Wash and brighten with dilute sulphuric acid.

Naphthalene Pink [$C_{30}H_{21}N_3 \cdot HCl + H_2O$].—This colouring matter, also called Magdala Red, and derived from amido-azo-naphthalene, is the hydrochloride of the base rosa-naphthylamine. It is but little used, namely, for the purpose of obtaining on silk bright pinks, which have a strong yellowish-red fluorescence.

Application to Silk.—Dye in a bath containing

"boiled-off" liquor, with or without the addition of sulphuric acid. Brighten with dilute sulphuric or tartaric acid. The colour is faster than that given by Magenta, Eosin, or Safranine; it is fast to dilute acids and alkalis, but not to light.

Safranine [$C_{21}H_{22}N_4 \cdot HCl$].—In chemical constitution this red colouring matter is apparently allied to Magenta, and is the hydrochloride of a colour-base safranine. It is applied to the various fibres in the same manner as Magenta. On wool the colour is not fast to light. Strictly speaking, the name Safranine is given to several closely-allied products. Fuchsia (Soc. Ch. Ind., Basle) is dimethyl-aniline-safranine.

Application to Cotton.—Prepare the cotton with tannic acid and tartar emetic, wash and dye in a neutral bath at 50° C. One may also steep the cotton in a solution of lead acetate (with or without previous impregnation with a solution of soap), dry, and dye in a neutral bath of the colouring matter; the colour thus obtained is objectionable because of the lead it contains. Fixed with tannic acid and tartar emetic the colour is fairly fast to light.

Application to Wool and Silk.—Dye in the same manner as with Magenta.

Neutral Red (L. Casella & Co.).—This colouring matter and others called Neutral Blue and Neutral Violet, being allied to Safranine, are all applied to the various textile fibres by similar methods. They yield dull shades of red, blue, and violet respectively, not fast to light on wool. They are of little use in wool- and silk-dyeing, but may be used with advantage by the cotton-dyer for producing compound shades.

New Blue D (Casella & Co.).—This colouring matter gives a colour closely resembling that of vat-indigo blue, which on cotton is extremely fast to light. Although affected by alkalis, it is well adapted for cotton-dyeing, and may in many cases replace vat-indigo blue. New Blue D is frequently used in conjunction with Methylene Blue or other basic colouring matters.

Application to Cotton.—Mordant the cotton with tannic acid and tartar emetic, and dye in a neutral bath of the colour solution.

(c.) Aniline Black Group.

Aniline Black. — Unlike other colouring matters, Aniline Black is not a commercial article. For the purpose of the dyer it must be produced upon the fibre itself. Little or nothing is known of its chemical constitution. It is a product of the oxidation of a salt of aniline, generally analine hydrochloride, and appears to exist in two states of oxidation. The less oxidised product is a blue-black, somewhat sensitive to the action of acids, particularly sulphurous acid, under the influence of which it acquires a greenish tint. The original colour can only be temporarily restored by treatment with an alkaline solution. The more highly oxidised product is a violet-toned black, which is not turned green by acids. This is produced by submitting the former to a supplementary oxidation. It is remarkable for its extreme fastness to acids, alkalis, light, etc., and is indeed one of the most permanent dyes known.

Application to Cotton.—For dyeing cotton Aniline Black the most usual oxidising agent employed is bichromate of potash or chromic acid. According to the temperature at which the dyeing is effected, two methods may be distinguished, namely, the warm method and the cold method.

Warm Method.—For 100 kg. = 220·4 lb. of cotton the dye-bath contains the following ingredients 1,600 litres = 352 gallons water, 40 kg. = 88·1 lb. of hydrochloric acid, 34° Tw. (Sp. Gr. 1·17), 10 kg. = 22·0 lb. of aniline, 10-14 kg. = 22·0-30·8 lb. of bichromate of potash.

These proportions may be varied according to the particular shade of black required. A portion of the hydrochloric acid may also be replaced by an equivalent amount of sulphuric acid. Use, for example, 24 kg. = 52·8 lb. of hydrochloric acid, and 4-6 kg. = 8·8-13·2 lb. of sulphuric acid, 168° Tw. (Sp. Gr. 1·84). The intensity of the colour, however, is always regulated by the amount of aniline employed.

The aniline and hydrochloric acid, diluted slightly with water, are carefully mixed in a suitable glazed earthenware vessel, and the acid solution of aniline hydrochloride thus obtained is added to the dye-bath previously filled with cold water. The bichromate of potash is dissolved separately in a little warm water and added to the bath.

ANILINE COLOURING MATTERS. 67

The cotton is worked for 1 hour in the cold solution, until, indeed, it has acquired a considerable intensity of colour, after which the temperature is gradually raised to 50°-60° C. The whole operation may last from 1-3 hours.

Another method is as follows: Dye the cotton in the cold for 1 hour with only half the quantity of the several ingredients added to the bath, then add the remainder, and continue the dyeing in the cold for 1 hour longer;

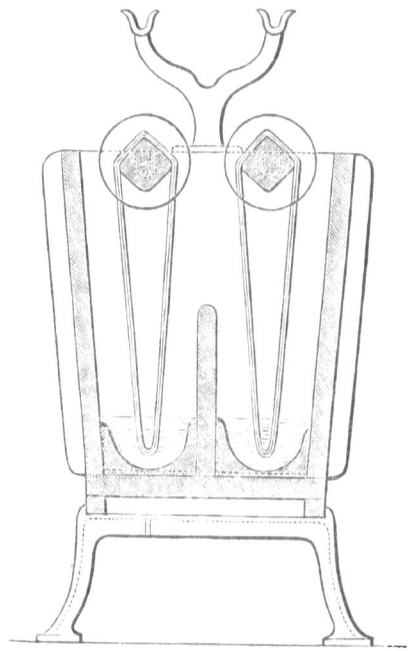

Fig. 5.—Aniline Black Dyeing Machine.

after this raise the temperature gradually to 50°-60° C., and continue the dyeing for another hour.

The more concentrated the solution and the greater its acidity, the more rapidly does the dyeing take place. Excess of acid and prolonged heating tend to give bronze-coloured blacks, and much of the colouring matter is only superficially fixed. If, however, the heating has been of short duration, the black has a bluish tone, and is liable to become green under the influence of acids.

68 FABRIC DYEING & TEXTILE COLORING MIXTURES.

It is essential that the temperature of the bath should be raised very gradually, otherwise there is a great loss of colouring matter, since much of it is then precipitated in the bath and not on the fibre.

After dyeing, the cotton must be well washed with water, then boiled in a solution of soap containing 5-10 g. = 77·1-154·3 grains per litre = 1 quart, with or without the addition of a little carbonate of soda, and finally dried.

Cold Method.—According to this method the dyeing operation is conducted entirely in the cold, the proportions of the ingredients and the concentration of the bath being altered to render this possible.

For 100 kg. = 220·4 lb. of cotton use 16-20 kg. = 35·2-44·0 lb. of hydrochloric acid, 20 kg. = 44·0 lb. of sulphuric acid, 8-10 kg. = 17·6-22·0 lb. of aniline, 14-20 kg. = 30·8-44·0 lb. of bichromate of potash, 10 kg. = 22·0 lb. of ferrous sulphate. The quantity of water should be very much smaller than in the warm method, otherwise the dyeing would either be incomplete or would take too long. Fig. 5 represents an apparatus for dyeing cotton yarn Aniline Black. It is specially designed so that the hanks can be properly manipulated in as little liquid as possible. It consists of a strong wooden dye-bath, about 2 m. = 6½ feet long, and longitudinally divided into two compartments, each with rounded bottom. Above are two corresponding square rollers, each capable of holding about 5 kg. = 11 lb. of cotton yarn; there is also a support with two arms, on which the rollers can be placed either at the end of the dyeing process or for the purpose of filling them with yarn before beginning. Several such dye-baths are arranged in line, the rollers being turned by power, alternately to right and left, in order to avoid entanglement of the hanks. The use of such a machine saves labour, prevents the corrosion of the workmen's hands by the chromic acid, and gives a more regularly oxidised black.

It will be noticed that the proportions of bichromate of potash and acid employed in the cold method are larger than in the warm method; this is in order to facilitate the oxidation of the aniline salt at the lower temperature. The addition of the sulphuric acid has a similar effect, but

ANILINE COLOURING MATTERS. 69

it also tends to yield ultimately a more pleasing tone of black. The use of hydrochloric acid produces blue-blacks, while sulphuric acid alone gives such as are of a reddish hue. The addition of the ferrous sulphate is for the purpose of rendering the black less liable to turn green; of course, in the bath it is changed to ferric sulphate, and this acts as an oxidising agent.

The method of preparing the dye-bath for the cold method is similar to that already described. The aniline hydrochloride is previously made by mixing the aniline and hydrochloric acid; separate solutions of the ferrous sulphate, bichromate of potash, and sulphuric acid are also kept in readiness. The cotton is first worked for about 1 hour with only half the full amount of the several ingredients in the bath. At the end of this time, when the cotton has already acquired quite a black colour, it is lifted out, the other half of the ingredients is added, the cotton is then re-entered, and the dyeing is continued 1-1½ hour longer.

After dyeing, the cotton is well washed and boiled with a solution of soap and carbonate of soda, as previously stated, and dried. The use of soap alone gives violet-toned blacks; the addition of carbonate of soda makes the shade bluer.

Although the black produced by either of the above methods, but especially by the cold method, is tolerably stable, it is necessary, in order to render it perfectly ungreenable, to submit the dyed cotton, after washing, to a supplementary oxidation. Several methods have been proposed for this purpose, but perhaps the following, in which ferric sulphate is the oxidising agent, is the most serviceable : Prepare a mixture of 20 kg. = 44·0 lb. of ferrous sulphate, 5 kg. = 11 lb. of bichromate of potash, 15-18 litres = 33·0-39·6 lb. of sulphuric acid, 168° Tw., 60-70 litres = 13·2-15·4 gals. of water. Add 5 litres = 1·1 lb. of this mixture to 500 litres = 110·0 lb. of water, and work the cotton in the solution for ¾ hour at 75° C. then wash well, boil with soap, and dry.

Other Methods.—Another method, depending for its efficacy upon the optical effect that a mixture of violet and green appears blue, is to dye the black in a weak solution of Methyl Violet. This violet is fixed, it is supposed, by reason of the cotton itself having been partially oxidised and changed into oxycellulose during the dyeing process.

Two other methods of producing Aniline Black on cotton, though not practically employed, possess sufficient interest to deserve mention. One is that borrowed from the method so successful in the printing of calico with Aniline Black. It is based upon oxidising the aniline salt by means of potassium chlorate in the presence of vanadium. The cotton is impregnated with a somewhat concentrated solution containing 5-20 % of aniline hydrochloride (according to the intensity and fastness of the black required), 2-10 % of potassium chlorate, and a very minute quantity of vanadium chloride (not more than $\frac{1}{300}$ of the weight of aniline hydrochloride employed). After wringing out the excess of liquid, the colour is developed by hanging the cotton in an ageing stove heated to 25°-30° C., and in which the air is kept slightly moist by admitting a little steam. The chief defect of this process is that an uneven colour is liable to be produced, since the oxidation will take place unequally if there is a partial drying of the fibres, or an unequal exposure of the fibres to the air.

The other method referred to is that proposed by Goppelsroeder, in which a vat of reduced Aniline Black is made, the cotton being dyed in it just as in an indigo-vat.

The Aniline Black is first prepared separately, namely, by heating a solution containing aniline hydrochloride, potassium chlorate, ammonium chloride, and copper sulphate. The black pigment thus produced is purified by boiling with water, and afterwards with alcohol. It is then heated with a solution of caustic potash, and the colour-base of the black thus liberated is washed, dried, and dissolved in fuming sulphuric acid. This solution is poured into cold water, and the greenish-black precipitate thus produced is dissolved in caustic alkali, and reduced by heating with the addition of glucose, hydrosulphite of soda, or zinc powder. Ferrous sulphate and lime are inoperative. If cotton be steeped in the brownish-yellow solution thus obtained, and then exposed to the air, it acquires gradually a blue colour. By submitting this colour to a supplementary oxidation it changes to a light

ANILINE COLOURING MATTERS. 71

grey or deep black, according to the concentration of the vat. A judicious combination of the aniline black vat with an indigo-vat may yield very fast deep blues.

Application to Wool and Silk.—These fibres, especially the former, cannot be dyed satisfactorily with Aniline Black. It would appear as if the reducing action of the fibres themselves hindered the oxidation of the aniline salt. Better results are said to be obtained if the fibres are previously oxidised by immersing them for some time either in a weak solution of permanganate of potash, or in a dilute solution of bleaching-powder to which hydrochloric acid has been added. After this preliminary process, the wool or silk is washed and dyed by a process exactly analogous to the one described for dyeing cotton.

Naphthylamine Violet.—This colouring matter, derived from naphthylamine hydrochloride in a manner similar to that by which Aniline Black is obtained, must also be produced upon the fibre itself.

(*d.*) *Aniline Colours Containing Sulphur.*

Aldehyde Green $[C_{22}H_{27}N_3S_2O]$.—This colouring matter, now seldom used, is invariably prepared by the dyer himself, by the action of aldehyde upon a solution of Magenta dissolved in strong sulphuric acid.

$$Methylene\ Blue.\quad N \begin{cases} C_6H_3 \cdot N(CH_3)_2 \\ >S \\ C_6H_3 \cdot N(CH_3)_2 \cdot Cl \end{cases}$$

Methylene Blue, a derivative of dimethyl-aniline, gives a greenish blue.

Application to Cotton.—The cotton is prepared with tannic acid and tartar emetic, then washed and dyed in a separate bath containing Methylene Blue solution. The addition to the dye-bath of a small quantity of carbonate of soda or ammonia is beneficial. Enter the cotton cold, and raise the temperature gradually to 100° C.

Application to Wool.—Add to the dye-bath the requisite amount of colour in solution (0·5-5 %) and 1-2 % of carbonate of soda (crystals) or ammonia. Enter the wool cold, and raise the temperature gradually to 100° C. in the course of ½ hour, and boil ½ hour longer.

Ethylene Blue (from diethyl-aniline) resembles Methylene Blue.

CHAPTER V.

QUINOLINE AND PHENOL COLOURING MATTERS.

Flavaniline (Meister, Lucius, and Brüning) [$C_9H_5N(CH_3)\cdot C_6H_4(NH_2)\cdot HCl$].
This basic yellow colouring matter, derived from acetanilide, is the hydrochloride of the quinoline base flavaniline. It is applied to cotton, wool, and silk, in the same manner as Magenta. On wool the colour is developed a little by passing the dyed material through dilute sulphuric acid. Brighter colours are obtained on wool mordanted with thiosulphate of soda according to the method employed when dyeing with Methyl Green. Flavaniline yellow is not fast to light.

Flavaniline S is an alkali salt of the sulphonic acid of the flavaniline base. It is applied to wool and silk in an acid bath.

Quinoline Blue [$C_{28}H_{35}N_2I$].—This disused colouring matter, also called Cyanine, is applied to cotton, wool, and silk in the same manner as Magenta. The colours it yields are very fugitive towards light.

Quinoline Yellow (Berlin Aniline Co).—This colouring matter is the sodium salt of the sulphonic acid of quinoline-phthaleïn. It gives a pure yellow colour, and is applied in an acid dye-bath in the same manner as other sulphonic acid colouring matters.

(*a.*) *Nitro Compounds of Phenol.*

Picric Acid [$C_6H_2(NO_2)_3\cdot OH$].—This colouring matter is tri-nitro-phenol; it is used only in silk- and wool-dyeing. Cotton has no attraction for it, and although it may be fixed on this fibre by means of albumen, the method has no practical value. The animal fibres, on the contrary, readily take up Picric Acid from an acid solution. It gives a clear bright yellow, free from any tinge of orange; indeed, when compared with most other yellows, it appears to have a greenish hue.

Application to Wool.—Dye for ½-1 hour, at 60°-100° C.,

QUINOLINE AND PHENOL COLOURING MATTERS. 73

with 1-4 % of Picric Acid, with the addition of 2-4 % of sulphuric acid, 168° Tw. (Sp. Gr. 1·84). The bath must be preserved, since it cannot be exhausted.

By repeated washing with water only is it possible to remove nearly the whole of the colour from the dyed fibre. It does not stand milling well, both on this account and because the colour becomes brownish under the influence of alkalis.

It is also not a good dye for the woollen part of mixed goods (wool and cotton), since it comes off in the tannin bath used for preparing the cotton. It is frequently employed for the production of compound colours, *e.g.* with Methyl Green and with Indigo Carmine for yellowish greens, with Acid Violet for olive, etc.

A noteworthy feature of Picric Acid yellow is that on exposure to light it rapidly becomes darker, acquiring a dull orange colour, which does not readily fade.

Application to Silk.—Dye with 0·5-1 % of Picric Acid in a bath slightly acidified with sulphuric acid, with or without the addition of " boiled-off " liquor. The colour marks off on paper, if submitted to pressure for some time. The compound colours dyed with Picric Acid and Methyl Violet are fluorescent.

The method of dyeing a weighted yellow on silk by first preparing the fibre with lead acetate, and then dyeing with Picric Acid, is not to be recommended, since the colour blackens in the presence of sulphuretted hydrogen, and the picrate of lead causes the silk, when ignited, to burn like touch-paper.

Phenyl Brown.—This colouring matter, which consists of a mixture of dinitrophenol $C_6H_3(NO_2)_2·OH$ with an amorphous brown substance, is not now used in dyeing to any very large extent. It is not applicable to cotton.

Application to Wool.—It gives nice brown shades, which are said to be very fast to light.

Dye in a bath slightly acidified with sulphuric acid. If the wool is boiled with bichromate of potash after dyeing, the colour assumes a redder tone.

Application to Silk.—Dye in a bath slightly acidified with sulphuric acid.

Victoria Yellow $[C_6H_2(CH_3)(NO_2)_2·OK]$.—This colouring matter is the potassium salt of dinitro-*p*-cresol. The dyes it yields are so very fugitive that it is now no longer employed. It is applied to wool and silk in the same way as Picric Acid.

Campobello Yellow [$C_{10}H_6(NO_2)\cdot ONa$].—This colouring matter is the sodium salt of a-nitro-a-naphthol; it was formerly also sold as French Yellow and Chryseïnic Acid. It is applied in the same way as Naphthol Yellow, and gives similar shades. It is applicable only to silk and wool, but the colour is neither fast to light nor to washing.

Naphthol Yellow [$C_{10}H_5(NO_2)_2\cdot ONa + H_2O$].—This colouring matter is the sodium or calcium salt of dinitro-a-naphthol; it is also known by the following names: Martius Yellow, Manchester Yellow, Golden Yellow, Saffron Yellow, Primrose, Naphthaline Yellow, etc. It is applicable only to silk and wool, but, since it is volatile even at low temperatures, it has the defect of marking off.

Application to Wool.—Dye in an acid bath for 1 hour with 0·5-3 % of colouring matter and 3-6 % of sulphuric acid, 168° Tw. (Sp. Gr. 1·84). Enter the wool at 40° C., and heat gradually to 100° C. If desirable, one may also acidify with 5-10 % of alum instead of sulphuric acid. According to the amount of colouring matter employed, the colour varies from a pale lemon-yellow to a deep and brilliant orange-yellow. It is not fast to light, and is not suitable for goods which have to be milled.

Application to Silk.—It is applied in the same manner as Picric Acid.

Naphthol Yellow S (BASF) [$C_{10}H_4\cdot(NO_2)_2\cdot ONa\cdot SO_3Na$]. —This colouring matter is the sodium salt of the sulphonic acid of Naphthol Yellow; it is used only in silk- and wool-dyeing. It is very much faster to washing than either Picric Acid or Naphthol Yellow. It is not volatile on steaming, and does not mark off, but is fugitive to light. It is applied in the same way as Naphthol Yellow.

New Yellow (Bayer & Co.) [$C_{10}H_5\cdot NO_2\cdot OK\cdot SO_3K$] is the potassium salt of the nitro derivative of β-naphthol-a-monosulphonic acid.

Application to Wool.—The method of dyeing is the same as with Naphthol Yellow, and similar shades are produced. The colour is not fast to light. The addition of a small percentage of stannic chloride to the dye-bath adds brilliancy to the colour.

Application to Silk.—Dye in a bath slightly acidified with sulphuric acid, and without the addition of "boiled-off" liquor or soap.

Palatine Orange (BASF) [$C_{12}H_4(NO_2)_4(ONH_4)_2$].—This colouring matter is the ammonium salt of tetra-nitro-γ-diphenol. It is applied to wool and silk in a bath acidulated with sulphuric or acetic acid.

Heliochrysin [$C_{10}H_3(NO_2)_4ONa$].—This colouring matter is the sodium salt of tetra-nitro-naphthol; it is also known as Sun Gold. It gives fine orange shades on wool and silk, but it is not fast to light, and has been little employed in practice.

Aurantia [$N(C_6H_2(NO_2)_3)_2NH_4$].—This orange colour-matter is the ammonium salt of hexa-nitro-diphenylamine; it is also known as Imperial Yellow, and is only applicable to wool and silk.

Dye in a bath very slightly acidified with sulphuric acid. Contact of the solution with metallic surfaces must be avoided, since these render the solution brown. Only glass or wooden vessels should be employed for dissolving or dyeing. It has been stated that Aurantia has decidedly poisonous properties, and occasions skin eruptions.

(*b.*) *Colouring Matters produced by the Action of Nitrous Acid on Phenols.*

Resorcin Blue [$C_{18}H_3(NH_4)Br_6N_2O_5$](?).—This colouring matter, also known as Fluorescent Blue, is the ammonium salt of hexa-brom-diazo-resorufin. It is applicable to wool and silk only, more particularly the latter.

Silk dyed with this colour is remarkable for its reddish fluorescence, the red colour appearing very prominently by gas-light. It is said to be fast to light, washing, and acids. When used in combination with other colouring matters it gives pleasing shades, all possessing fluorescence.

Application to Silk.—Dye in a bath containing "boiled-off" liquor or soap, and neutralised with acetic acid. Brighten in a cold bath slightly acidified with tartaric or sulphuric acid.

Naphthol Green (L. Casella & Co.) [$C_{10}H_5O_7SNFe$](?). —This colouring matter is the iron compound of nitroso-naphthol-mono-sulphonic acid. It gives an olive-green colour on wool, remarkable for its fastness to light. The colour also bears the action of milling with soap fairly well, but it is much impoverished by the action of carbonate of soda. Naphthol Green is not applicable in cotton-dyeing.

Application to Wool.—Dye at 100° C., with the addition of 2-3 % of sulphuric acid, 168° Tw. (Sp. Gr. 1·04), 20 % of sodium sulphate, and 10 % of ferrous sulphate. It is very useful for producing compound shades in conjunction with other colouring matters which are applied in an acid bath.

(c.) *Rosolic Acid Colours.*

$$\text{Aurin.} \quad \text{C} \begin{cases} C_6H_4\cdot OH \\ C_6H_4\cdot OH \\ C_6H_4\cdot O \end{cases}$$

This colouring matter in its commercial form is also called Yellow Corallin. Red Corallin or Peony Red is produced from Yellow Corallin by the action of ammonia on it at high temperatures. Owing to the extreme fugitiveness to light, soap, and acids of the colours they yield, these colouring matters are seldom employed in dyeing. Silk and wool may be dyed in a weak soap bath, heating it gradually to the boiling point; brighten the silk with tartaric acid. They are still used by the calico and woollen printer.

(d.) *Phthaleïns.*

$$\text{Fluoresceïn.} \quad \text{C} \begin{cases} C_6H_3\cdot OH \\ C_6H_3\cdot OH \\ C_6H_4\cdot CO\cdot O \end{cases} O$$

Fluoresceïn dyes wool and silk yellow in a slightly acidulated bath; but it is scarcely used in dyeing, because it does not give fast colours, and these are surpassed in beauty of tint by those yielded by other yellow colouring matters. Its sodium compound, known under the name of Uranin, is applied in the same manner.

$$\text{Chrysolin.} \quad \text{C} \begin{cases} C_6H_3\cdot O(C_7H_7) \\ C_6H_3\cdot ONa \\ C_6H_4\cdot CO\cdot O \end{cases} O$$

This colouring matter is the sodium salt of benzyl-fluoresceïn. It dyes wool and silk in a neutral bath, although a better result is obtained by previously mor-

QUINOLINE AND PHENOL COLOURING MATTERS. 77

danting with alum. The yellow colour produced is similar to that given by Turmeric, but it is said to be faster to light. In cotton-dyeing it may be used for topping Quercitron Bark yellow.

Eosins.—There are quite a number of red colouring matters belonging to the class of Eosins. They differ from each other both in their composition and the shade they produce.

Eosin J (soluble in water). $C \begin{Bmatrix} C_6HBr_2 \cdot OK \\ C_6HBr_2 \cdot OK \\ C_6H_4 \cdot CO \cdot O \end{Bmatrix} O$

This is the potassium salt of tetra-brom-fluoresceïn. It dyes a yellowish-pink shade.

Eosin B (soluble in water) is an alkali salt of tetra-iodo-fluoresceïn, and dyes a bluish-pink shade. It is also known by the names Erythrosin, Pyrosin B (P. Monnet), and Soluble Primrose (Durand and Huguenin).

Aureosin (Willm, B. & Girard) is a chlorinated fluoresceïn.

Rubeosin (Willm, B. & Girard) is a nitro-chlor-fluoresceïn.

Eosin BN (BASF), also called Safrosin (Soc. Chem. Ind., Basle), is a brom-nitro-fluoresceïn.

Lutécienne (Poirrier) is a mixture of brom-nitro-fluoresceïn, with di- and tetra-nitro-fluoresceïn or Poirrier's Orange 2.

Nopalin and Imperial Red contain dinitro-naphthol mixed with brom-nitro-fluoresceïn.

Coccin is a mixture of Aurantia with brom-nitro-fluoresceïn.

Eosin (soluble in alcohol) is the potassium salt of tetra-brom-fluoresceïn-methyl or ethyl-ether. It bears also such names as Methyl-eosin, Ethyl-eosin, Primrose (soluble in alcohol), Rose JB, Erythrin, Methyl-erythrin, etc.

The methyl compounds gives yellower shades than the ethyl compound. Both give better and brighter shades than the ordinary eosins, which they have largely displaced in silk-dyeing.

Rose Bengal is the sodium salt of tetra-iodo-dichlor-fluoresceïn.

Phloxin (P. Monnet & Co.) is the potassium salt of tetra-brom-dichlor-fluoresceïn.

Cyanosin (P. Monnet & Co.) is the potassium salt of the methyl-ether of Phloxin.

The eosins may be used in cotton-, wool-, or silk-dyeing, and yield bright shades, which vary from yellowish-scarlet to bluish-crimson. The yellowest shade is given by Eosin J or G, then follow Methyl and Ethyl-eosin, Eosin B, Phloxin, and Safrosin. The bluest shade is given by Rose Bengal. Cyanosin, Phloxin, and others are sold, however, in different shades, and are marked accordingly with J, R, or B. Safrosin is not quite so bright on silk as some of the other blue shades, but on wool it gives a good full and bright colour.

The eosins soluble in alcohol give brighter shades than those soluble in water, but, apart from the cost of the alcohol required, they cannot always be used, because the strong yellowish fluorescence which they give is sometimes not desirable.

Application of Eosins to Cotton.—Cotton has naturally no attraction for the eosins, but since they form lakes with metallic oxides (especially protoxides), it is possible to dye this fibre if it is previously mordanted with some metallic salt. The mordants employed are those of zinc and lead. Aluminium salts are also used.

Impregnate the cotton with a cold or tepid solution of sulphated oil, 100 g. = 3·2 ozs. per litre = 1 quart; wring out, dry and steam; then work in aluminium acetate, at 8° Tw. (Sp. Gr. 1·04) for ½ hour, steep 1-2 hours longer, and wring out. Dye in a fresh bath of Eosin solution acidified with 5-10 % of alum. Enter the cotton at 45° C., and allow the bath to cool gradually during the dyeing process. A strong hot solution of soap may be substituted for the sulphated oil, and a solution of nitrate acetate or basic acetate of lead at 5° Tw. (Sp. Gr. 1·025) may replace the aluminium acetate. The colours thus obtained have a yellow tone.

Very good bluish shades are produced by impregnating the cotton with the lead solution, then drying and dyeing in Eosin solution.

The use of a lead salt has the disadvantage that the colour is blackened if exposed to an atmosphere containing sulphuretted hydrogen. The dye thus obtained is naturally of a poisonous character.

QUINOLINE AND PHENOL COLOURING MATTERS. 79

Application to Wool.—Dye with 0·5-2 % of Eosin, with the addition of 5-10 % of alum. Enter the wool at 40° C., and raise the temperature gradually to 100° C. in the course of 1 hour. If the bath is acidified with acetic or sulphuric acid instead of alum, the shades produced are not so bright, but the wool is less harsh. With Erythrosin the temperature of the dye-bath must be kept below the boiling point.

The water used should be as free from lime as possible, otherwise it causes precipitation and loss of colouring matter. If lime is present, neutralise it with acetic acid before dyeing.

Application to Silk.—Dye in a bath slightly acidified with acetic or sulphuric acid, with or without the addition of "boiled-off" liquor. Wash and brighten with acetic or tartaric acid. Add the colour solution gradually, and heat slowly to 60° C.

$$\text{Galle\"in.} \quad C \begin{cases} C_6H_2 \begin{cases} OH \\ O \end{cases} \\ C_6H_2 \begin{cases} O \\ OH \end{cases} \\ C_6H_4 \cdot CO \cdot O \end{cases}$$

This fine purple colouring matter, sometimes called Anthracene Violet, is derived from phthalic anhydride and pyrogallol. It is sold in the form of a reddish-brown powder or a 10 % paste, not very soluble in cold water, but readily so in hot. It gives fine purple shades on cotton, wool, and silk, which are tolerably fast both to light and soap.

Application to Cotton.—Prepare the cotton with aluminium, chromium, or iron mordants in the usual manner, and dye in a separate bath with Galleïn. The whole process is identical with that used in dyeing with Alizarin, Logwood, or other polygenetic colouring matters.

All the mordants yield purple colours, those obtained by the use of chromium and iron being bluish, those of tin reddish, and those of aluminium intermediate in tone. All the colours may be regarded as fast to light and soap.

Application to Wool.—Mordant the wool with 2 % of bichromate of potash. The addition of sulphuric acid, even to the extent of 1 %, is injurious, and dulls the

FABRIC DYEING & TEXTILE COLORING MIXTURES.

colour. Dye in a separate bath with 10-20% of Galleïn paste, containing 10 % of solid matter. Enter the wool cold, and raise the temperature gradually to the boiling point. The shade thus produced is bluish-purple or violet.

With aluminium mordant a much redder and brighter purple is given. Mordant with 6-8 % of aluminium sulphate and 5-7 % of cream of tartar. With the addition of 1-2 % of acetate of lime (solid), the shade is somewhat more intense and slightly brighter. The addition of chalk to the dye-bath is not to be recomended; even with 2 % the colour is much deteriorated. With iron mordants Galleïn gives a deep violet colour. Use 8 % ferrous sulphate and 5 % tartar. The single-bath method is also applicable.

All the above Galleïn colours are specially adapted for goods which have to be milled. The chromium mordant is the most generally useful of those mentioned.

Silk is dyed in the same way as with Alizarin.

$$Cærule\ddot{\imath}n \quad C_6H_4 \begin{cases} COC_6H_2 & -O \\ & >O \\ COC_6H(OH) \cdot O \end{cases}$$

This green colouring matter, also called Anthracene Green, is derived from Galleïn by the action of sulphuric acid at a high temperature. It is sold in two forms, either as a thick black paste (Cœruleïn paste) containing 10-20 % of Cœruleïn, or as a black powder. The former is more or less insoluble in water, and requires in some cases the addition of bisulphite of soda to render it soluble. The latter, known as Cœruleïn S, is soluble in water, and is indeed a compound of Cœruleïn with bisulphite of soda $[C_{20}H_8O_6 \cdot 2NaHSO_3]$; this form is the one most easily applied. Cœruleïn is mostly employed in calico-printing for producing very fast olive-green shades. The colours it yields both on cotton and on wool are remarkable for their fastness to light, acids, alkalis, etc., and whenever its price permits, it will find an extensive use in the dyeing of these materials. Whatever the mordant used, only different shades of olive-green are obtained. The use of copper dye-vessels should be avoided.

Application to Cotton.—If insoluble Cœruleïn is heated with a mixture of caustic alkali (NH_3) and zinc powder,

QUINOLINE AND PHENOL COLOURING MATTERS. 81

a brownish-red solution of the reduction product Cœruleïn is obtained, which, on exposure to air, immediately becomes green again, with precipitation of the original Cœruleïn. This brownish-red liquid, or "Cœruleïn-vat," as it might be termed, may be used for dyeing after the manner of the Indigo-vat.

If the soluble Cœruleïn S is employed, the cotton must be previously prepared with aluminium, chromium, iron, or tin mordant, according to the usual methods, and then dyed in a simple solution of the colouring matter. Care should be taken to begin dyeing at a low temperature, and to raise it very gradually to 100° C. During the dyeing process sulphurous acid is given off, and the liquid becomes green and alkaline. The water employed should be free from salts of lime and other alkaline earths, since these produce insoluble lakes with Cœruleïn. The insoluble form of Cœruleïn may be applied in the same way, if bisulphite of soda is added to the bath to render it soluble, but the results are not quite so satisfactory.

Application to Wool.—Mordant with 2-3 % bichromate of potash and 0-0·7 sulphuric acid, 168° Tw. (Sp. Gr. 1·84). Without sulphuric acid the colour is slightly paler. Dye in a separate bath containing only Cœruleïn S. Enter the wool cold, and raise the temperature very gradually (say in the course of ½ hour) to 60° C. Dye at this temperature for about 1 hour, then heat gradually in the course of ½ hour to 100° C., and boil for ¼ hour. The addition to the dye-bath, during the last ¼ hour, of 1-2 % of chalk, makes the shade bluer, but generally speaking the addition of chalk or calcium acetate to the bath is to be avoided. With· 2 % of Cœruleïn S a pale sage-green is obtained, with 5 % a medium olive-green, and with 10 % a very dark green, almost black. These colours may be used instead of Indigo-greens, being equally fast to light, milling, etc.

With aluminium mordant shades can be obtained which are somewhat bluer or greyer than with bichromate of potash, but they are very apt to be uneven.

With iron mordant dirty olive and olive-black shades are obtained. Use 4 % of ferrous sulphate and 8 % of cream of tartar, and dye with 0·5-10 % of Cœruleïn S.

If wool is mordanted with an amount of pure stannic

F

chloride equivalent to 5 % $SnCl_2·2H_2O$ (tin-crystals), it needs the addition of 40 % of cream of tartar to yield a normal bluish-green colour, when dyed afterwards with 5 % of Cœruleïn S; but it is remarkable that even without the addition of any tartar a full greyish-black colour is obtained. (With the majority of polygenetic colouring matters, stannic chloride is an unsatisfactory mordant.) With 5 % of Cœruleïn S, the colour is perhaps too much like a bad black to be of general use, but with 0·5-2 % very pleasing greys are obtained.

Application to Silk.—Cœruleïn has scarcely been introduced into silk-dyeing, though it is capable of giving good fast shades. Mordant in the usual manner with alum, dye in a separate bath with Cœruleïn S, and brighten with a solution of soap.

(*e.*) *Indophenols.*

a-Naphthol Blue (L. Casella & Co.).

$$C_6H_4 \begin{cases} N(CH_3)_2 \\ N = C_{10}H_5(OH) \end{cases}$$

Dimethyl-amido-phenyl-oxy-naphthylamine.

—This colouring matter, also called Indophenol Blue N, is produced by oxidising a mixture of dimethyl-*p*-phenylenediamine and α-naphthol, or by the action of nitrosodimethyl-aniline on α-naphthol. It gives colours very similar to vat-indigo blues, and which are said to be moderately fast to light. They are, however, extremely sensitive to the action of acids; even weak acids destroy the blue colour and change it to brown. Indophenol Blue N is better adapted for woollen- and calico-printing than for dyeing.

Under the influence of reducing agents—*e.g.* glucose and caustic soda, stannous chloride and carbonate of soda, etc.—Indophenol Blue is changed into Indophenol-White, which is soluble in pure or acidulated water.

For the preparation of indophenol-white, mix together 10 kg. = 22·0 lb. of Indophenol Blue (paste) and 60 litres = 13·2 gals. of water; add 30 litres = 6·6 gals. of a 10 % solution of tin crystals ($SnCl_2·2H_2O$), and heat to 25° C. until reduction takes place.

Application to Cotton.—Dye for 10 minutes at 40° C. in a solution containing 5-10 g. = 77·1-154·3 grains of

QUINOLINE AND PHENOL COLOURING MATTERS. 83

indophenol-white per litre, then wring out and wash, and develop the colour by working the cotton for about 2 minutes at 50° C. in a dilute solution of bichromate of potash. Better colours are obtained if the cotton is previously prepared with sulphated oil.

Application to Wool.—Dye for 15 minutes at 80° C. in a solution of indophenol-white, rendered either alkaline by the addition of sodium carbonate, or acid by means of acetic acid. Wring out, wash, and develop the colour by exposure to air, or by working the material for a few minutes in cold dilute solution of bichromate of potash, or an ammoniacal solution of sulphate of copper. For dark shades the solution of indophenol-white should be concentrated; the dye-bath is not exhausted, and should always be preserved.

Gallocyanin (Durand and Huguenin).—This colouring matter, also called New Fast Violet, is obtained by the action of nitroso-dimethyl-aniline on tannic acid. In dyeing, it yields a bluish-violet colour possessing only moderate brilliancy, but tolerably fast to the action of acids, alkalis, and light. Applied in conjunction with other colouring matters, it is useful for obtaining compound shades.

Application to Cotton.—Mordant the cotton by means of an alkaline solution of chromium oxide, and wash well. Dye in a separate bath with Gallocyanin, at a temperature of 80° C., for 1-1½ hour. If, after dyeing, the cotton is washed, dried, and steamed, the colour becomes somewhat darker and faster.

Application to Wool.—The wool may be dyed without any previous preparation, or it may be first mordanted in the usual manner with bichromate of potash. Dye in a neutral bath. Introduce the wool into a cold solution, raise the temperature gradually to 70° C. in the course of 1 hour, and continue dyeing for ½-1 hour longer.

Application to Silk.—Dye at 70°-80° C. in a bath containing colour solution and "boiled-off" liquor. The silk may be previously mordanted with chrome alum, though this is not absolutely necessary.

CHAPTER VI.

AZO COLOURING MATTERS.

(a.) Amido-azo-colours.

Aniline Yellow [$C_6H_5 \cdot N = N \cdot C_6H_4 \cdot NH_2 \cdot HCl$].—This
Diamido-azo-benzene-hydrochloride.
colouring matter is no longer used in dyeing, because the colour which it yields is volatile and not fast. Cotton has no attraction for it. Wool and silk may be dyed in a bath slightly acidified with acetic acid.

Chrysoïdine [$C_6H_5 \cdot N = N \cdot C_6H_3(NH_2)_2 \cdot HCl$].—This
Diamido-azo-benzene-hydrochlor d .
colouring matter, much used in cotton-dyeing for producing various shades of orange, is prepared by the action of diazo-benzene-chloride on *m*-phenylene-diamine. It is well adapted for shading, and may be used as the yellow part in a number of compound shades—*e.g.* olive, brown, scarlet, etc. Chrysoïdine FF (L. Casella & Co.) is the corresponding toluene compound.

Application to Cotton.—Mordant the cotton with tannic acid and tartar emetic, and wash; dye at 60° C. in a solution of the colouring matter, without any further addition. Avoid high temperatures, since the colour is thereby rendered duller.

Sometimes the fixing of the tannic acid with tartar emetic may be omitted, and for very light shades it is not even necessary to prepare the cotton with tannic acid, since this fibre seems to possess naturally a certain attraction for Chrysoïdine. Good shades are obtained by applying Chrysoïdine to cotton previously dyed with Catechu, Sumach, etc.

Application to Wool.—Dye at 60°-70° C. in a neutral bath, or with the addition of 2-4 % of soap, or one acidified with alum. These additions tend to brighten the colour. The addition of sulphuric acid to the dye-bath impoverishes the colour, but if, after dyeing according to the above method, the wool be worked for 10-15 minutes in cold water slightly acidified with sulphuric acid, the

AZO COLOURING MATTERS. 85

colour acquires a deeper and redder hue. Dyeing at 100° C. dulls the colour considerably.

Application to Silk.—Dye at a temperature of 60° C., with or without the addition of a little soap to the dyebath. Brighten in a bath very slightly acidified with sulphuric acid.

Phenylene Brown.—
$$[C_6H_4\cdot(NH_2)\cdot N = N\cdot C_6H_3(NH_2)_2\cdot 2HCl].$$
Triamido-azo-benzene-hydrochloride.

This colouring matter is prepared by the action of nitrous acid on *m*-phenylene-diamine, and dissolving the base thus produced in hydrochloric acid. It also bears the commercial names: Bismarck Brown, Vesuvine, Canelle, Manchester Brown, Cinnamon Brown, etc. Bismarck Brown GG and EE.(L. Casella & Co.) are the pure products of toluylene-diamine and phenylene-diamine respectively.

Application to Cotton.—Prepare the cotton with tannic acid and tartar emetic; wash and dye in a neutral bath at 50°-60° C. Add the colour solution gradually. A slight addition of alum to the dye-bath may sometimes be made to modify the shade. The shades given by Bismarck Brown are similar to those obtained from Catechu, but, as a rule, brighter. Light shades can be dyed without previous preparation of the cotton. Catechu browns are frequently dyed with it in order to brighten or modify the colour.

A great variety of compound colours are obtainable by using it along with other basic colouring matters—*e.g.* Magenta, Malachite Green, Methyl Violet, Methylene Blue, etc.

Application to Wool.—Dye in a neutral bath. For a full shade, use 5-8 % of colouring matter. The addition of 8-10 % of alum to the bath makes the shade redder. Enter the wool at 45° C., and heat gradually to 100° C.

Application to Silk.—Dye in a weak soap bath at 60° C., and brighten in a fresh bath slightly acidified with acetic or tartaric acid.

(*b.*) *Amido-azo-sulphonic Acids.*

Fast Yellow.—
$$[(SO_3Na)C_6H_4\cdot N = N\cdot C_6H_4(NH_2)].$$
Amido-azo-benzene-sodium-p-sulphonate.

This colouring matter is also called Acid Yellow; it cannot be used for dyeing cotton. It is well adapted for using along with other acid colouring matters to obtain compound shades on wool or silk. Employed alone, it cannot compete with some other yellows in brilliancy. The above compound is sometimes distinguished as Fast Yellow G, while Fast Yellow R is given to the corresponding toluene compound.

Application to Wool.—Dye in an acid bath. For 0·5-3 % of colouring matter add 2-6 % of sulphuric acid, 168° Tw. (Sp. Gr. 1·84). Enter the wool at 40° C., and heat gradually to 100° C. in the course of ¾-1 hour, and boil for ¼ hour. If 5-10 % of alum be used instead of sulphuric acid, the shade given is weaker and less orange.

Application to Silk.—Dye at a temperature of 60°-80° C., in a bath containing "boiled-off" liquor and acidified with sulphuric acid.

Dimethylaniline Orange.—

$$[(SO_3 \cdot NH_4)C_6H_4 \cdot N = N \cdot C_6H_4(N(CH_3)_2)].$$
p-Dimethyl-amido-azo-benzene-ammonium-p-sulphonate.

Other commercial names of this colouring matter are Helianthin, Gold Orange, Orange III., and Tropæolin D, etc.

Application to Cotton.—Work the cotton in cold stannate of soda solution, 5° Tw. (Sp. Gr. 1·025), till thoroughly saturated, and wring out; work for ¼-½ hour in a cold solution of alum (15-20 %), and wring out; dye in a concentrated solution of the colouring matter, with the addition of an equal percentage of alum. Enter cold, and heat gradually to 45° C., but not higher. Dry without previous washing. The colour is not fast to washing.

Application to Wool.—Dye exactly as with Fast Yellow. Somewhat brighter shades are obtained by using stannic chloride instead of sulphuric acid. With 2 % of colouring matter a full reddish-orange is obtained.

Application to Silk.—Dye exactly as with Fast Yellow.

Diphenylamine Orange.—

$$[(\overset{4}{SO_3}K)C_6H_4 \cdot N = N \cdot C_6H_4(N \cdot H \cdot C_6H_5)]$$
p-Phenyl-amido-azo-benzene-potassium-p-sulphonate.

This colouring matter is also called Tropæolin OO,

AZO COLOURING MATTERS.

Orange IV., Orange N, Yellow N, etc. It is very sensitive to the action of an excess of free acid, which causes it to dye a more orange colour. Large excess of mineral acid causes its solutions to become violet through liberation of the free colour acid. Closely allied to this colouring matter are the three following:—

Metanil Yellow (BASF).—

$$[(SO_3Na)C_6H_4 \cdot N = N \cdot C_6H_4(N \cdot H \cdot C_6H_5)]$$
p-Phenyl-amido-azo-benzene-sodium-*m*-sulphonate.

This colouring matter is also called Tropæolin G (L. Cassella & Co.).

Brilliant Yellow (BASF).—
$$[(SO_3Na)C_6H_3(CH_3) \cdot N = N \cdot C_6H_4(N \cdot H \cdot C_6H_5)]$$
p-Phenyl-amido-azo-toluene sodium-*p*-sulphonate.

Azoflavin 2 (BASF).—
$$[(SO_3Na)C_6H_4 \cdot N = N \cdot C_6H_4(N \cdot H \cdot C_6H_4(NO_2))]$$
p-Nitro-phenyl-amido-azo-benzene-sodium-*p*-sulphonate.

All these colouring matters are specially suitable for wool- and silk-dyeing, and give fine yellow or orange shades. They are applied in the same way as Fast Yellow and Dimethylanine Orange. The colours on cotton are not fast to washing. If 10 % of alum is added to the dye-bath instead of sulphuric acid, the colours on wool are rendered brighter.

Indian Yellow (L. Casella & Co.) is isomeric with Azoflavin.

Congo Red (Berlin Aniline Co.).—

$$\begin{cases} C_6H_4 \cdot N = N \cdot C_{10}H_5(NH_2)(SO_3Na) \\ C_6H_4 \cdot N = N \cdot C_{10}H_5(NH_2)(SO_3Na) \end{cases}$$
Tetrazo-diphenyl-dinaphthylamine-sodium-disulphonate.

This colouring matter possesses the very interesting property of being readily applied to the vegetable fibres without the aid of a mordant. It may be used for dyeing mixed goods consisting of cotton and wool, and yields a bright scarlet colour, fairly fast to boiling soap solutions, but not to light. It is also extremely sensitive to the action of acids; these change the colour to blue.

Application to Cotton.—Dye at 100° C. in a neutral bath, or one rendered slightly alkaline by the addition of soap; wash and dry. A much richer colour is got if the cotton is previously mordanted with stannic oxide, or, better still, with sulphated oil and aluminium sulphate.

Application to Wool and Silk.—Dye in a neutral bath or with the addition of a little soap.

Benzopurpurin (Bayer & Co.).—

$$\begin{cases} C_7H_6 \cdot N = N \cdot C_{10}H_5(NH_2)(SO_3Na) \\ | \\ C_7H_6 \cdot N = N \cdot C_{10}H_5(NH_2)(SO_3Na) \end{cases}$$
Tetrazo-ditolyl-diphenylamine-sodium-disulphonate.

This colouring matter is applied to the various fibres in the same manner as Congo Red, being closely allied to it in chemical constitution. It yields a bright scarlet colour, fairly fast to soap, and less sensitive to light and particularly to acids, than Congo Red. It is not affected by dilute acetic acid, or even by dilute mineral acids. The best addition to make to the dye-bath is phosphate of soda, though one may, if desirable, use soap or silicate of soda instead.

(*c.*) *Oxy-azo Colouring Matters.*

These include yellow, orange, red, crimson, and brown colours. Nearly all belong to the class of so-called acid-colours, and are specially suitable for wool- and silk-dyeing. When applied to cotton, most of the colours are not fast to washing with water.

Many of the scarlets have largely displaced Cochineal in wool-dyeing. For plain scarlet dyes (on flannels, etc.) they are even preferable to Cochineal, since the colour does not become dull and bluish on washing with soap. They are, however, not suitable for yarn which has to be woven with other light-coloured yarns, if the material so produced must afterwards be washed with soap, scoured, or milled. During these processes the colour "bleeds," or comes off slightly and dyes very permanently the contiguous light-coloured fibres, thus spoiling the general appearance of the fabric. This defect is common, indeed, to all those coal-tar colouring matters which dye without mordant. For dark-coloured fabrics the defect is not noticeable.

The dyeing properties of many of these colouring matters are very similar.

AZO COLOURING MATTERS.

The dyes they yield are very fairly fast to light, though they differ considerably in this respect. As a rule, the tetrazo compounds are faster to light than the diazo compounds.

It is difficult to identify all the colouring matters of this class met with in commerce, since each manufacturer gives a special name and mark to his own products. The following list, however, gives a selection :—

Tropæolin Y.—

$$[(SO_3Na)C_6H_4\cdot N = N\cdot C_6H_4\cdot(OH)]$$
p-Phenol-azo-benzene-sodium-*p* sulphonate.

This colouring matter has now little importance, having been replaced by other similar but superior colouring matters.

Resorcinol Yellow.—

$$[(SO_3Na)C_6H_4\cdot N = N\cdot C_6H_3(OH_2)]$$
Resorcinol-azo-benzene-sodium-*p*-sulphonate.

This colouring matter is also called Tropæolin O, Tropæolin R, Chryséolin, Chrysoïn (Poirrier). It gives an orange dye of only moderate brilliancy, and is chiefly used along with other acid-colours to produce olives, maroons, etc.

Orchil Brown (Bayer & Co.).—

$$[(NH_2)C_{10}H_6\cdot N = N\cdot C_6H_4(SO_3Na)].$$

This colouring matter is *a*-naphthylamine-azo-benzene-sodium-sulphonate.

Azarin S (M. L. & B.).—

$$C_6H_2Cl_2 \begin{cases} OH \\ N = N\cdot C_{10}H_7(OH) \\ | \\ SO_3NH_4\cdot H \end{cases}$$

This colouring matter is suitable only for cotton. It yields a brilliant red, said to be fairly fast to light.

The following method of applying it to calico is proposed by the manufacturers :—

Pad the cloth with a solution of aluminium acetate to which a small proportion of stannous oxide has been added. After drying and ageing for 12 hours, work the cloth for ½ hour in a cold solution containing a very small proportion of acetate of lime and sodium carbonate; then wash well and dye in a solution of the colouring matter, with the addition of a little sulphated oil. After dyeing, wash, dry, pad with a dilute solution of sulphated oil, dry and steam; then wash in a cold weak soap solution and dry. The alkalinity of the soap gives the red a bluish tone, which may be removed by a final passage in very dilute acid.

Azo Blue (Bayer & Co.).—

$$\begin{cases} C_7H_6 \cdot N = N \cdot C_{10}H_5(OH)(SO_3K) \\ | \\ C_7H_6 \cdot N = N \cdot C_{10}H_5(OH)(SO_3K) \end{cases}$$

Tetrazo-ditolyl-β-napthol-potassium-disulphonate.

This was the first blue colouring matter of the azo series, which possessed the property of dyeing vegetable fibres without the aid of a mordant. It yields a reddish-blue colour, fast to soap and concentrated mineral acids, and moderately fast to light.

It is applied in dyeing in the same manner as Benzopurpurin and Chrysamin, with which it may therefore be used in the same bath for the purpose of obtaining compound shades on cotton mixed goods, etc.

Chrysamin (Bayer & Co.).—

$$\begin{cases} C_6H_4 \cdot N = N \cdot C_6H_3(OH)(COONa) \\ | \\ C_6H_4 \cdot N = N \cdot C_6H_3(OH)(COONa) \end{cases}$$

Tetrazo-diphenyl diphenol-sodium-α-carbonate.

This colouring matter dyes the vegetable fibres without the aid of a mordant in the same manner as Congo Red, Benzopurpurin, and Azo Blue.

The colour obtained on cotton by the aid of a boiling soap-bath is a sulphur-yellow remarkably fast to light; also fast to acetic acid, but not to mineral acids.

α-Naphthol Orange.—

$$[(SO_3Na)C_6H_4 \cdot N = N \cdot \alpha C_{10}H_6 \cdot OH)]$$

α-Naphthol-azo-benzene-sodium-*p*-sulphonate.

This colouring matter is also called Tropæolin OOO No. 1, Orange No. 1 (Poirrier). It dyes a very reddish-orange shade.

β-Napthol Orange.—

$$[(SO_3Na)C_6H_4 \cdot N = N \cdot \beta C_{10}H_6 \cdot OH)]$$

β-Naphthol-azo-benzene-sodium-*p*-sulphonate.

This colouring matter also bears the names: Tropæolin OOO No. 2, Orange No. 2 (BASF) (Poirrier), Orange extra

AZO COLOURING MATTERS.

(L. Casella & Co.), Mandarin S (Berlin Aniline Co.), Chrysaureïn. It dyes a bright reddish-orange shade, similar to that yielded by Tropæolin OOO No. 1. A mixture of Tropæolin OOO No. 2 and Fast Red is sometimes sold under the name of French Red (*rouge Français*). Tropæolin OOO No. 2 is largely used, both alone and along with Indigo Carmine and other acid-colours, for producing browns, olives, etc.

Trapæolin OOOO.—

$$[C_6H_5 \cdot N = N \cdot \beta C_{10}H_5(OH)(SO_3Na)]$$
Benzene-azo-β-naphthol-sodium-α-sulphonate.

Orange G (M. L. & B.).—

$$[C_6H_5 \cdot N = N \cdot \beta C_{10}H_4(OH)(SO_3Na)_2]$$
Benzene-azo-β-naphthol-sodium-β-disulphonate.

This colouring matter dyes a bright orange, somewhat less red than Tropæolin OOO No. 2, which is extremely fast to light.

Scarlet G T (Bayer & Co.).—

$$[C_6H_4(CH_3) \cdot N = N \cdot \beta C_{10}H_5(OH)(SO_3Na)]$$
Toluene-azo-β-naphthol-sodium-β-sulphonate.

Xylidine Scarlet G (M. L. & B.).—

$$[C_6H_3(CH_3)_2 \cdot N = N \cdot C_{10}H_5(OH)(SO_3Na)]$$
o-Xylene-azo-β-naphthol-sodium-β-sulphonate.

Scarlet R (Bayer & Co.) is the isomeric *m*-xylene compound.

This colouring matter seems to be identical with Scarlet 2 R (Berlin Aniline Co.). Scarlet G (BASF) is the closely-allied *p*-xylene-azo-β-naphthol-sodium-α-disulphonate.

Xylidine Scarlet R (M. L. & B.).—This colouring matter is isomeric with the last, and is the sodium salt of the corresponding β-disulphonic acid compound. It is sometimes simply called Scarlet R. As the mark (R) implies, it dyes a redder shade than Scarlet G. Scarlet R (BASF) is the isomeric *p-m*-xylene-azo-β naphthol-sodium-disulphonate.

Scarlet G G (M. L. & B.).—

$$[C_6H_2(CH_3)_3 \cdot N = N \cdot C_{10}H_4(OH)(SO_3Na)_2]$$
Cumene-azo-β-naphthol-sodium-β-disulphonate.

This colouring matter seems to be identical or isomeric with Scarlet 4 R (Berlin Aniline Co.).

Scarlet R R.—This colouring matter is isomeric with the last, and is the sodium salt of the corresponding a-disulphonic acid.

Another very closely allied colouring matter, also called Scarlet R R, is the following

$$[C_6H_2\cdot(CH_3)_2(C_2H_5)\cdot N = N\cdot C_{10}H_4(OH)(SO_3Na)_2]$$
Ethyl-xylene-azo-β-naphthol-sodium-a-disulphonate.

Scarlet 3 R (M. L. & B.).—

$$[C_6H_2(C_2H_5)(CH_3)_2\cdot N = N\cdot C_{10}H_4(OH)(SO_3Na)_2]$$
Ethyl-dimethyl-azo-β-naphthol-sodium-disulphonate.

Scarlet 3 R (BASF).—

$$[C_6H_2(CH_3)_3\cdot N = N\cdot C_{10}H_4(OH)(SO_3Na)_2]$$
Pseudo-cumene-azo-β-naphthol-sodium-disulphonate.

Scarlet 4 R (M. L. & B.).—

$$[C_6H_2(CH_3)_3\cdot N = N\cdot C_{10}H_4(OH)(SO_3Na)_2]$$
Cumene-azo-β-naphthol-sodium-disulphonate.

Fast Brown (M. L. & B.).—

$$[(SO_3Na)C_6H_2(CH_3)_2\cdot N = N\cdot aC_{10}H_6(OH)]$$
a-Naphthol-azo-xylene-sodium sulphonate.

Fast Red (BASF).—

$$[(SO_3Na)C_{10}H_6\cdot N = N\cdot \beta C_{10}H_6(OH)]$$
β-Naphthol-azo-naphthalene-sodium-sulphonate.

This colouring matter is also called Roccellin (Poirrier), Orseillin No. 3, Rubidin, Rauracienne, etc. It does not yield a brilliant red, but one which is remarkable for its fullness or body. It is very useful in conjunction with other acid-colours for producing compound shades.

Fast Brown (BASF).—This colouring matter is the corresponding a-naphthol compound.

Fast Red C (BASF).—

$$[(NH_2)C_{10}H_6\cdot N = N\cdot aC_{10}H_5(OH)(SO_3Na)]$$
a-Naphthylamine-azo-a-naphthol-sodium-sulphonate.

Croceïn 3 BX (Bayer & Co.).—

$$[(SO_3Na)C_{10}H_6\cdot N = NC_{10}H_5(OH)(S\overset{a}{O_3}Na)]$$
Naphthalene-sodium-sulphonate-azo-β-naphthol-sodium-a-sulphonate.

Claret Red B (M. L. & B.).—

$$[C_{10}H_7\cdot N = N\cdot C_{10}H_4(OH)(S\overset{a}{O_3}Na)]$$
a-Naphthalene-azo-β-naphthol-sodium-a-disulphonate.

This colouring matter is also called Bordeaux R. If

alum is used in the dye-bath for wool, the colour is apt to be very uneven.

Fast Red B (BASF).—This colouring matter is isomeric with the last, being the sodium salt of the corresponding β-sulphonic acid. Other names which seem to be given to it are Claret Red R (M. L. & B.) and Bordeaux G.

Crystal Scarlet 6 R (L. Casella & Co.).—This colouring matter is also isomeric with Claret Red B, being the sodium salt of the corresponding γ-disulphonic acid.

Amaranth (M. L. & B.) also (L. Casella & Co.).—

$$[(SO_3Na)C_{10}H_6 \cdot N = N \cdot C_{10}H_4(OH)(SO_3Na)_2]$$
Naphthalene-azo-β-naphthol-sodium-trisulphonate.

New Coccin (M. L. & B.); *Fast Red D* (BASF).—These colouring matters are isomeric with the last.

Brilliant Scarlet 4 R (L. Casella & Co.).—This colouring matter is also isomeric with Amaranth, being the sodium salt of the corresponding γ-disulphonic acid.

Scarlet 6 R (M. L. & B.).—

$$[(SO_3Na)C_{10}H_6 \cdot N = N \cdot C_{10}H_3(OH)(SO_3Na)_3]$$
Naphthalene-azo β-naphthol-sodium-tetrasulphonate.

Anisol Red (BASF).—

$$[C_6H_4(OCH_3) \cdot N = N \cdot C_{10}H_5(OH)(SO_3Na)]$$
Anisol-azo-β-naphthol-sodium-sulphonate.

Phenetol Red (BASF).—

$$[C_6H_4(OC_2H_5) \cdot N = N \cdot C_{10}H_4(OH)(\overset{a}{SO_3Na})_2]$$
Phenetol-azo-β-naphthol-sodium-α-disulphonate.

This colouring matter is also called Cocinin (M. L. & B.).

Coccinin B (M. L. & B.).—This colouring matter is closely allied to the last, being the corresponding methyl-*p*-cresyl $C_6H_3(OCH_3)(CH_3)$ compound.

Brilliant Croceïn M (L. Casella & Co.).—

$$[C_6H_5 \cdot N = N \cdot C_6H_4 \cdot N = N \cdot C_{10}H_4(OH)(SO_3Na)_2]$$
Benzene-azo-benzene-azo-β-naphthol-sodium-γ-disulphonate.

Scarlet S (Berlin Aniline Co.).—This colouring matter is isomeric with the last.

Scarlet 5 R (M. L. & B.).—

$$[C_6H_5 \cdot N = N \cdot C_6H_4 \cdot N = N \cdot C_{10}H_3(OH)(SO_3Na_3)]$$
Benzene-azo-benzene-azo-β-naphthol-sodium-trisulphonate.

Biebrich Scarlet (Kalle & Co.).—

$$[(SO_3Na)C_6H_4 \cdot N = N \cdot C_6H_3(SO_3Na) \cdot N = N \cdot C_{10}H_5(OH)]$$
β-Naphthol-azo-benzene-azo-benzene-sodium-disulphonate.

This colouring matter is identical with Imperial Scarlet (Bayer & Co.).

Fast Scarlet (BASF).—

$(SO_3Na)C_6H_4 \cdot N = N \cdot C_6H_2(SO_3Na) \cdot N = N \cdot C_{10}H_5(OH)]$
β-Naphthol-azo-benzene-azo-benzene-sodium-sulphonate.

Crocein Scarlet 3 B (Bayer & Co.).—

$[(\overset{4}{SO_3}Na)C_6H_4 \cdot N = N \cdot C_6H_4 \cdot N = N \cdot C_{10}H_5(OH)(\overset{a}{SO_3Na})]$
Benzene-sodium-*p*-sulphonate-azo-benzene-azo-β-naphthol-sodium-*a*-sulphonate.

Crocein Scarlet 7 B (Bayer & Co.).—

$[(SO_3Na)C_6H_3(CH_3)N = N \cdot C_6H_3(CH_3) \cdot N = N \cdot C_{10}H_5(OH)(SO_3Na)]$
Toluene-sodium-*p*-sulphonate-azo-toluene-azo-β-naphthol-sodium-*a*-sulphonate.

Scarlet S S (Berlin Aniline Co.).—This colouring matter is isomeric with the last. A mixture of Xylidine Scarlet and Acid Magenta is said to have been sold under the same name.

Application of the Oxy-azo Colours to Cotton.—Work the cotton ¼-½ hour in a cold solution of stannate of soda, 6°-8° Tw. (Sp. Gr. 1·03-1·04), wring out and pass into a cold solution of alum, 4°-6° Tw. (Sp. Gr. 1·02-1·03) for ¼ hour. Wring out, and, without washing, dye at 45°-50° C., in a concentrated solution of the colouring matter, with the addition of 5-10 % of alum. Wring out, and dry without washing. The colour does not withstand even washing with water. The cotton may also be previously prepared with sulphated oil, instead of with stannic oxide, and then dyed in the manner described.

An interesting method of obtaining faster colours on cotton with these colouring matters is that included in the patents of Grässler and others.

The colouring matters used for wool and silk are the alkali salts of complex sulphonic acids, and are produced in a cold alkaline solution by the mutual reaction of azo compounds upon phenols.

The colouring matter obtained is soluble if either the azo compound or the phenol, or both, are in the form of sulphonates, but if otherwise, it is insoluble.

The insoluble non-sulphated colour can be precipitated at once upon the cotton fibre, by first impregnating the latter with a cold alkaline solution of the phenol,

AZO COLOURING MATTERS.

wringing out, and then passing it into a solution of the azo compound. In practice, it is desirable to pass again into the phenol solution, wring out, wash and soap slightly, in order to remove loosely adhering colour.

Somewhat brighter and fuller colours are also obtained by preparing the cotton previously with sulphated oil. Owing to its instability, the solution of the azo compound should be prepared only a short time before use.

The dye produced on the cotton in this way bears washing with water, and even with soap solutions, but it is liable to rub off; it also withstands the action of light fairly well.

The following equation represents the formation of Xylidine Red in this way :—

$$C_6H_3(CH_3)_2 \cdot N = N \cdot Cl + C_{10}H_7 \cdot ONa =$$
Diazo-xylene-chloride.   Sodium-α-naphthol.

$$C_6H_3(CH_3)_2 \cdot N = N \cdot C_{10}H_6 \cdot OH + NaCl.$$
Xylidine red.

Application to Wool.—Dye with 1-2 % of colouring matter, with the addition of 2-4 % of sulphuric acid, 168° Tw. (Sp. Gr. 1·84), and 15-30 % of sodium sulphate. Enter the wool at 40°-50° C., raise the temperature gradually, in the course of 1 hour, to 100° C., and boil ¼ hour.

Several of the colouring matters give brighter colours if the sulphuric acid is replaced by 5-10 % of alum, or 5-10 % of stannic chloride, 120° Tw. (Sp. Gr. 1·6). Care must always be taken to have the bath sufficiently acid to develop the full colouring power; and if there is any tendency to uneven dyeing, the temperature should be raised very gradually.

Application to Silk.—Dye in a bath containing "boiled-off" liquor, slightly acidified with sulphuric acid.

The above modes of application do not, of course, apply to Azarin S, Azo Blue, and Chrysamin.

CHAPTER VII.

ANTHRACENE COLOURING MATTERS.

Alizarin [$C_{14}H_6O_2(OH)_2$]. — This valuable colouring matter, formerly known as a substance only obtainable from Madder root, is now made in large quantities from the coal-tar product, anthracene.

Alizarin is the best type of those colouring matters which dye only with the aid of a mordant, and which yield various colours according to the mordant employed (polygenetic colouring matters). In itself it has little or no colouring power, having no affinity for the vegetable fibres, and merely imparting a comparatively fugitive orange-brown colour to the animal fibres. It possesses, however, the valuable property of forming variously-coloured insoluble precipitates or lakes when combined with many of the metallic oxides, and it is on this property that its use in dyeing depends. Its compound with alumina is red, with stannous oxide orange, with chromic oxide claret-brown, and with ferric oxide violet. All the colours produced on the textile fibres by means of these mordants are extremely fast to light, boiling with soap solutions, etc.

Very closely allied to Alizarin are the colouring matters Isopurpurin or Anthrapurpurin, Flavopurpurin, and Purpurin, $C_{14}H_5O_2(OH)_3$. Their method of application is so similar to that employed for Alizarin that only special reference will be made to each where points of difference arise. They are sold separately or mixed together in various proportions, each manufacturer giving his own brand to the different qualities and mixtures. It is customary, for the sake of simplicity, to sell them, whether separate or mixed, under the common name, "Alizarin."

Those which consist entirely, or most largely, of Alizarin are called the blue shades of Alizarin, while those in which Flavopurpurin or Isopurpurin predominate, constitute the yellow shades of Alizarin. These designations have arisen because the former dye alumina mordanted cotton a crimson or bluish shade of red, while the latter give a scarlet or yellow shade of red.

Application to Cotton.—Alizarin serves principally for the production of the brilliant Turkey-red dye, already referred to under the head of Madder. For this purpose it has entirely supplanted Madder and its commercial preparation Garancin, because the colours it yields are far more brilliant, quite as fast, and less expensive.

Turkey-red dyeing probably had its origin in India. At an early date it was introduced into Turkey (hence its name), and about the middle of the eighteenth century it began to be practised in France.

Since the publication of the process in the year 1765 by the French Government, it has been carried on largely in Switzerland, Germany, and Britain. At the present time, the chief seats of this important industry are the Vale of Leven, near Glasgow, and Elberfeld, in Germany.

Numerous alterations and improvements have been gradually introduced, until it has now reached a very high state of perfection indeed. All the details of the process now employed have been empirically determined throughout a long period, and the successful production of the best Turkey-red depends upon their careful execution.

Cotton is dyed Turkey-red in the form of yarn and cloth. The process for yarn-dyeing seems to have experienced little change since the time when Madder was the dyestuff employed, and may serve as a type of the older methods of Turkey-red dyeing. It may be distinguished as the Emulsion process.

The present method of Turkey-red cloth-dyeing differs considerably in the earlier stages from that in vogue for yarn, and is known as Steiner's process (from the name of its inventor). There is, however, a third process of Turkey-red dyeing applicable to both cloth and yarn, which represents the method most recently introduced. This may be termed the sulphated-oil process.

Emulsion Process for Dyeing 500 $kg.$ = 1102·3 $lb.$ *of Turkey-red Yarn.*—The grey yarn is first " laced," *i.e.* the skeins constituting each " head " or " hank " are loosely fastened together by the intertwining of a short piece of cotton cord, in order to prevent entanglement during the several operations. The ends of the cord are also knotted, once or several times, for the purpose of subsequently recognising the various lots.

1st Operation: Boiling.—Boil the yarn 6-8 hours with a solution of carbonate of soda, 1° Tw. (Sp. Gr. 1·005).

Fig. 6.—Turkey-red Yarn-wringing Machine.

then wash well with water, squeeze and dry in a stove at 55°-60° C.

2nd Operation: First green liquor.—This liquor is an emulsion made up with 75 kg. = 165·3 lb. of olive oil, 8 kg. = 17·6 lb. of sheep-dung, about 1,000 litres = 220·0 gals. of water, and a sufficiency of a concentrated solution of carbonate of soda to make the whole to 2° Tw. (Sp. Gr. 1·01).

Work the hanks of yarn separately in this emulsion at

a temperature of 30°-40° C., till thoroughly saturated (about ½ minute), and wring out as evenly as possible. This process is usually called "tramping" (Fr., *tremper*, to steep).

Fig. 6 represents the taking-off end of a wringing machine made by Messrs. Duncan Stewart & Co., Glasgow, and used in some Turkey-red works in this country.

It consists of two large discs revolving on a common shaft. On the periphery of each, and directly opposite each other, are several large iron hooks connected with springs, toothed wheels, and rackwork, in such a manner that those on one of the discs are capable of twisting, while at the same time both sets of hooks yield inwards as the two discs revolve. The hanks of yarn are properly steeped in the emulsion by hand, and at once placed on a pair of the hooks; as the discs make a quarter of a revolution, the hooks twist and squeeze out the excess of liquor; during the next quarter of the revolution the hooks untwist themselves, and at the opposite side of the machine the hanks are thrown or pushed off by a pair of strong upright arms.

Fig. 7 represents a tramping machine of A. Weser, Elberfeld, and used in Germany, which performs the steeping as well as the wringing, hand-labour being required only for putting on and taking off the hanks of yarn. It consists essentially of the liquor trough E, above which are situated the fixed revolving roller B, and loose roller A, on which the hanks are suspended. D is an L-shaped arm, the horizontal portion of which passes within the loop of the hank and depresses it into the liquor. C is an iron cylinder pressing against B, and serves to impregnate the yarn with the solution.

The various movements of the machine are regular and automatic. The hank of yarn is placed on the rollers A and B when the arm D is in the horizontal position; the arm D at once falls and steeps the yarn in the liquid, and the rollers revolve for a short period; the arm D now again takes the horizontal position, the roller B ceases to revolve, and the roller A first twists, then untwists the hank, and it is ready to be removed by the attendant.

Whichever of the above machines is used, the work of "tramping" the yarn is practically continuous.

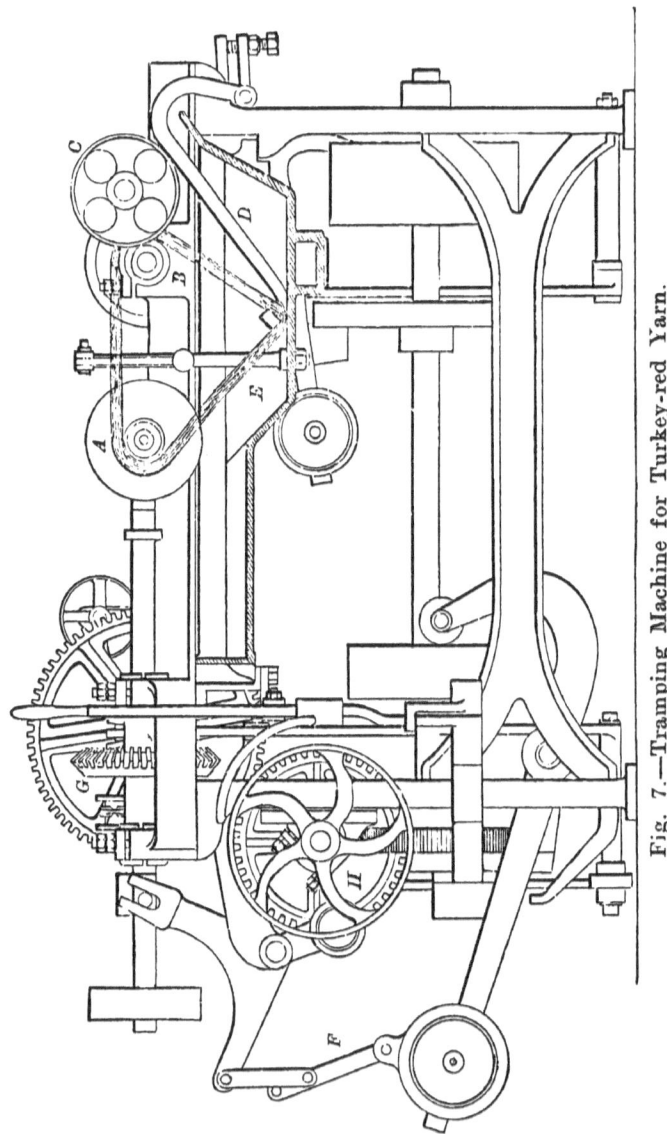

Fig. 7.—Tramping Machine for Turkey-red Yarn.

ANTHRACENE COLOURING MATTERS. 101

The prepared hanks are allowed to remain piled together over-night (12-20 hours), and are then dried in the stove. In this operation (stoving) the temperature is raised gradually to 55°-60° C., which is maintained for 2 hours. Care must be taken to allow the escape of the steam which is given off during the first stages of drying, otherwise the yarn is apt to be tendered.

3rd and 4th Operations: Second and third green liquors.—These are almost exact repetitions of the second operation, the liquor employed being made up separately and with the same proportions of the several ingredients as given above. The sole difference is that it is not necessary to let the prepared yarn lie in pile over-night; instead of this, if it is not raining, it is suspended on tin rods, and exposed to the open air for about 2-4 hours previous to stoving.

It is evident that, after stoving, the dry yarn is charged with sodium carbonate, and since it is very important that all the liquors should be maintained regularly at the same specific gravity, it is customary not to allow the liquor expressed during the wringing of the hanks to flow back into the " tramping " box, except in the case of the " first green liquor," but to collect it separately, and then, if necessary, to dilute it with water before using again.

The total amount of oil used is about 30 % of the weight of yarn, but only a portion of this becomes fixed on the fibre.

5th, 6th, 7th, and 8th Operations: First, second, third, and fourth white liquors.—The solution here used is simply carbonate of soda, at 2° Tw. (Sp. Gr. 1·01), but after working the yarn in it a short time, it necessarily becomes an oil emulsion from the oil stripped off the cotton, apart from the fact that it is always mixed with the surplus and expressed liquor from the similar operations with previous lots of yarn.

The yarn is " tramped " in the liquor, wrung out, exposed in the open air, and dried in the stove, as in the previous operations.

9th Operation: Steeping.—Steep the yarn during 20-24 hours in water heated to 55° C wash well, and dry in the stove at about 60° C. If the yarn contains much unmodified oil, a solution of carbonate of soda, at $\frac{1}{2}°$ Tw. (Sp. Gr. 1·0025), may be used; in this case a second steeping for 2 hours in tepid water is requisite before washing, etc.

10th Operation: Sumaching.—A decoction of Sumach is made by boiling 60 kg. = 132·2 lb. of best leaf Sumach for about $\frac{1}{2}$ hour, with sufficient water to make the cold filtered solution stand at $1\frac{1}{2}°$ Tw. (Sp. Gr. 1·0075). The stoved yarn, while still warm, is steeped in large vats in this decoction, as hot (40°-50° C.) as it can be borne by the boys who usually tramp it with bare feet beneath the surface of the solution. After steeping about 4-6 hours, the solution is drained off, and the excess is removed by a hydro-extractor.

11th Operation: Mordanting or Aluming.—A basic solution of alum is made by dissolving ordinary rock-alum in hot water, and when nearly cold, adding gradually a cold solution of one-fourth its weight of carbonate of soda crystals. The solution is made to stand at 8° Tw. (Sp. Gr. 1·04). Sometimes, though this is not essential, a further addition is made of about 150-200 cm^3. = 9·1-12·2 cubic inches of " red liquor," 16° Tw. (Sp. Gr. 1·08), and 5-7 g. = 77-108 grains troy of tin crystals ($SnCl_2$) per kg = 2·2 lb. of alum. The sumached yarn, while still damp, is tramped in the alum solution at a temperature of 40°-50° C., and left to steep for 24 hours. It is then thoroughly washed and hydro-extracted.

12th Operation: Dyeing.—Dye with 150-180 g. = 4·8-5·7 oz. of Alizarin (10 %), 30 g. = 462·9 grains of ground Sumach, and about 300 g. = 9·6 oz. of bullock's blood, per kg. = 2·2 lb. of cotton yarn. If the water contains little or no lime, add also ground chalk in the proportion of 1 % of the weight of Alizarin (10 %) employed. The yarn is introduced into the cold solution of the dye-vessel, the temperature is gradually raised to 100 C., in the course of 1 hour, and the boiling is continued for $\frac{1}{2}$-1 hour. After dyeing, the yarn is washed, although this is not absolutely necessary.

13th Operation: First Clearing.—Boil the yarn for 4 hours, at 3-4 pounds' pressure, with about 30 g. = 463·9 grains of carbonate of soda crystals, and 30 g. = 462·9 grains of palmeoil soap, dissolved in a sufficiency of water, per kg. = 2·2 lb. of yarn. Wash afterwards. The "clearing boiler" used, and shown in Fig. 8, and in plan in Fig 9, is similar in construction to an ordinary low-pressure bleaching kier; it is, however, made of copper instead of iron. A represents the yarn; B the lid provided with safety-valve and blow-off pipe; C the perforated false bottom; D the puffer-pipe; E the bonnet for distributing

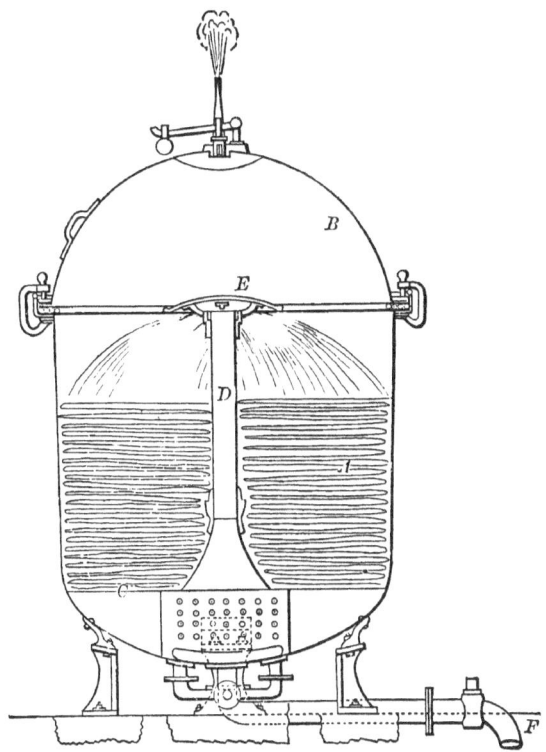

Fig. 8.—Clearing Boiler—Elevation.

the liquor over the yarn; F the draw-off pipe. During the boiling, the liquor which collects below the false bottom is forced by the steam up to the top of the puffer-pipe, there to be ejected and spread over the goods. This action is of an intermittent character, since, after each ejection of the liquor, the pressure of the steam must accumulate below the false bottom until it is again able to overcome the weight of the column of water in the puffer-pipe.

14th Operation: Second Clearing.—Boil the yarn for 1-2 hours at 3-4 pounds' pressure with a solution containing 25 g. = 385·1 grains of palm-oil soap and 1½ g. = 23·1 grains of tin-crystals per kg. = 2·2 lb. of yarn. Wash well and dry in an open-air shed. Previous to drying, the large excess of water is removed by means of the hydraulic press represented in Fig. 10. It consists of a strong iron framework D D, with a strong, fixed, but adjustable head A above, and a similar one B below, attached to the hydraulic piston C, and thus capable of being moved up or down. By means of this machine a very

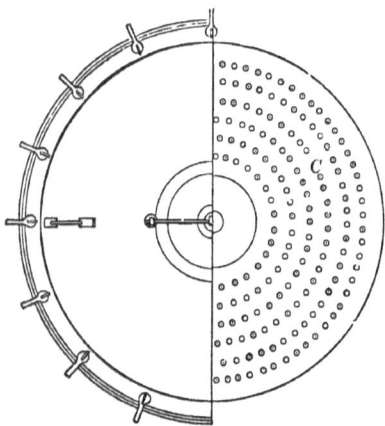

Fig. 9.—Plan of Clearing Boiler.

large quantity of wet yarn may be rapidly and efficiently squeezed.

The above fairly represents the "Emulsion process" of Turkey-red yarn-dyeing as practised at the present time. It consists, therefore, of a somewhat numerous series of operations, occupying usually about three weeks' time, and although, hitherto, no absolutely satisfactory scientific explanation has been given of the exact nature of the chemical changes effected by every detail of the whole process, still their general character is tolerably well understood. The object of the frequent steeping in oil-emulsion, drying in the open air, and stoving, is to impregnate the fibre evenly and thoroughly with oil, and to modify it in such a manner that it is not affected or removed by weak alkaline solutions, and that it will attract alumina from its solutions.

Many kinds of oil have been employed, but long

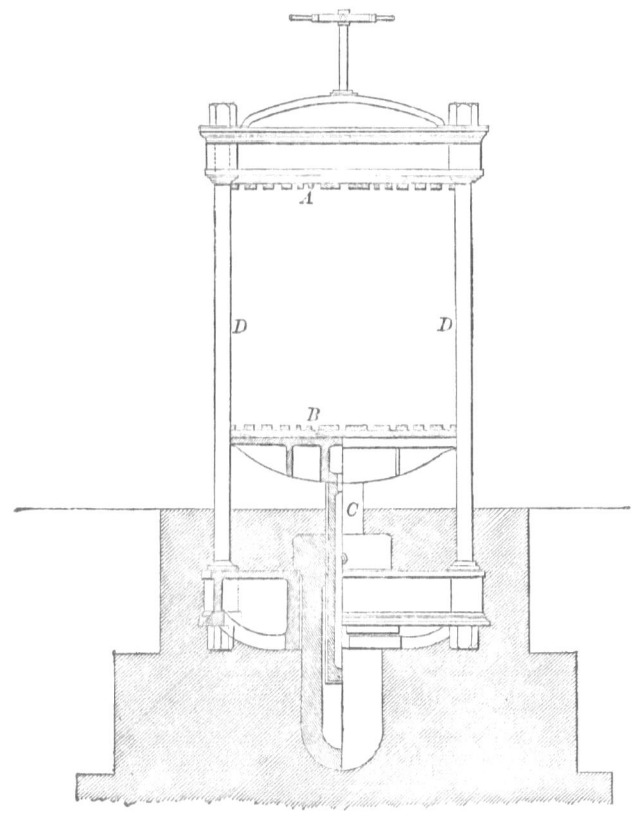

Fig. 10.—Hydraulic Press.

experience has proved that olive oil gives the best and most certain results. The particular quality of oil most suitable for the purpose is that obtained by a second pressing of the olives after they have somewhat fermented and been steeped in boiling water. (Fr., *huile tournate*). It contains nitrogenous and extractive matters, which cause it gradually to become rancid, particularly when exposed to the air, *i.e.* it decomposes, and a portion of the glycerine and fatty acids (margaric and oleïc acids) is liberated. One of the chief characteristics of a good olive oil suitable for Turkey-red is that, when 1 measure of it is shaken up with about 16 measures of sodium or potassium carbonate solution, at $3°$ Tw. (Sp. Gr. 1·015), it forms a white milky liquid or emulsion, from which the oil does not readily separate even after standing for 12-18 hours. The oil which forms the most perfect and permanent emulsion with the least quantity of potash or soda is the best. This property of emulsifying, however, can be readily imparted to any oil by mixing it with 5-15 % of oleïc acid.

The exact nature of the chemical changes which the oil undergoes during exposure to the air and stoving is unknown. It is probable, however, that, under the influence of the alkaline carbonate and heat, the oil is decomposed and oxidised in such a manner that there remains on the fibre essentially an insoluble oxyoleïc acid. Whatever may be the exact chemical composition of the modified oil, it has the property of fixing or combining with alumina, and the compound thus produced can further combine with Alizarin to form a red-lake. The effect which it has of giving brilliancy and fastness to the ultimate colour is probably, in part at least, due to its physical action of enveloping the coloured lake with a transparent oily varnish, which protects it more or less from external influences. All unchanged oil must be removed before mordanting (see Operation 9).

The impregnation of the cotton with tannin matter fixes an additional amount of alumina on the fibre, and tends to give deeper and fuller shades. Its use is, however, by no means absolutely essential, as seemed to be the case when Garancin or Madder was used, and by some dyers it is not used.

During the steeping in alum solution an insoluble basic aluminium compound is formed with the modified oil and also with the tannic acid if present. The complex mordant thus fixed on the cloth at this stage combines with Alizarin in the subsequent dye-bath to form the Turkey-red lake. The bullock's blood used is said to

ANTHRACENE COLOURING MATTERS.

prevent, by reason of the coagulation of its albumen, certain impurities accompanying the Alizarin from being fixed on the cotton, but some practical Turkey-red dyers say that blood-albumen, glue, and other substitutes which have been tried, cannot entirely replace it. It certainly adds brilliancy and purity to the colour.

The "first clearing" operation is for the purpose of removing any remaining impurities which the mordant may have attracted in the dye-bath, but for which its affinity is far less than for Alizarin.

The "second clearing" is said by some to introduce into the already extremely complex coloured lake a small portion of stannous oxide. Others allege that there is simply a tin-oleate produced, which is melted and spread over the fibre, as it were, without entering into chemical combination with the red-lake. Liechti has proved analytically that as much as 60 % of the fatty acid of the soap employed may disappear and become fixed in this manner upon the fibre.

The practical object of this operation is to give the colour the maximum purity and brilliancy of which it is capable.

Steiner's Process for Dyeing 500 $kg.$ = 9·8 *cwt. of Turkey-red Cloth.*—The main difference between this and the "emulsion process," already described, resides in the mode of applying the oil. In the process now to be described the cloth is impregnated with the requisite amount of oil at one operation, namely, by padding it in clear hot oil instead of in an oil-emulsion, after which it receives several passages through weak solutions of alkaline carbonate.

This method is capable of yielding a Turkey-red dye of exceptional brilliancy and intensity—better, indeed, than it is possible to obtain by the "emulsion process."

1st Operation: Bleaching.—The pieces are well washed and boiled, during 2-3 hours, with water only; then boiled for 10-12 hours with 22 litres = 4·8 gals. of caustic soda, 70° Tw. (Sp Gr. 1·35), and washed: then boiled a second time, for 10 hours, with 16 litres = 3·5 gals. of caustic soda, 70° Tw., and washed; and finally steeped for 2 hours in sulphuric acid, 2° Tw. (Sp. Gr. 1·01), well washed and dried.

108 FABRIC DYEING & TEXTILE COLORING MIXTURES.

In order to avoid tendering the fibre in the next operation, by reason of traces of acid left in the cloth, it is padded with carbonate of soda solution, at 4° Tw. (Sp. Gr. 1·02), and then dried.

2nd Operation: Oiling.—The cloth is padded in the open width in olive oil maintained at a constant temperature of 110° C.

Fig. 11 represents a section of the oil-padding machine of Messrs. Duncan Stewart & Co. It consists of a double-jacketed tank B (inside copper, outside iron) for contain-

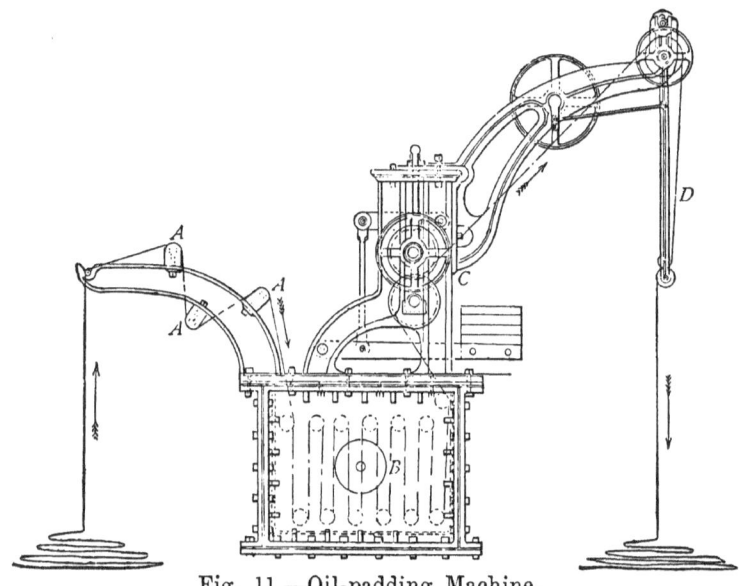

Fig. 11.—Oil-padding Machine.

ing the oil. It is heated by means of steam, and is provided with a series of rollers at the top and bottom. Above is a pair of heavy squeezing rollers C. The cloth is passed through as indicated in the figure, being well opened out and made free from creases before entering the oil, by means of the straining bars A A A, and afterwards loosely plaited down by the folder D.

After padding, the cloth is detached in ten-piece lengths and hung in the drying stove, the temperature of which is raised as rapidly as possible to 70° C., and this is maintained for 2 hours.

ANTHRACENE COLOURING MATTERS. 109

3rd to 9th Operation: Liquoring.—Pad the cloth seven times in the open width through a solution of carbonate of soda, at 4° Tw., and hang in the stove after each padding operation, maintaining the temperature in each case for 2 hours at 75°-77° C.

In winter the padding liquors are made warm (35°-40° C.), but in summer they are always cold, since if too hot, oil is stripped off the piece to an excessive and injurious degree. In the course of regular working, the liquors soon become veritable oil-emulsions, and constant

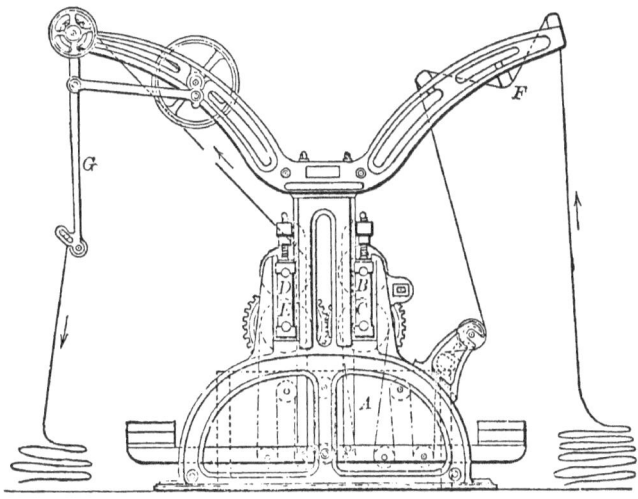

Fig. 12.—Section of Liquor-padding Machine.

oversight is necessary in order to maintain their specific gravity as constant as possible, and thus ensure ultimately a regular and satisfactory colour.

A section of the liquor-padding machine of Messrs. Duncan Stewart & Co. is shown in Fig. 12. It consists of a wooden box or tank A to hold the liquor, provided with rollers above and below. Over this are supported two pairs of heavy squeezing rollers B C and D E. At F a few straining bars serve to open out and stretch the cloth; G is the folder. The mode of passing the pieces through the machine is readily understood from the diagram.

With regard to the stoving, it is well to bear in mind

110 *FABRIC DYEING & TEXTILE COLORING MIXTURES.*

that during the first stages of drying much vapour is given off, and special attention must be given to ensure adequate ventilation. Fig. 13 is the ground plan, showing heating flues, and Fig. 14 is a sectional elevation of a modern four-storeyed Turkey-red stove. A A represent ordinary coal fires situated in the basement; the hot flue-gases

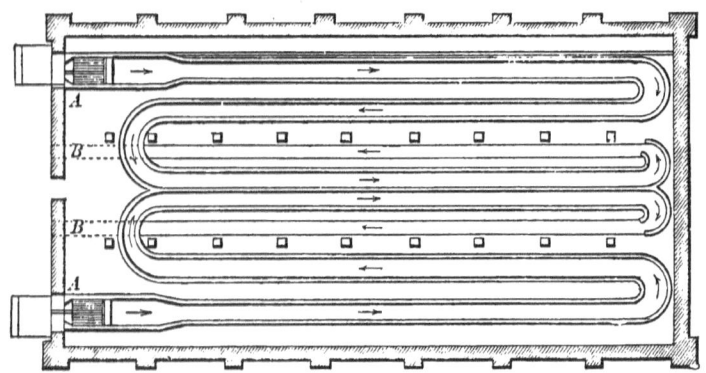

Fig. 13.—Plan of Turkey-red Stove.

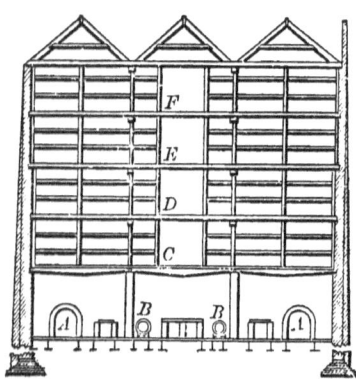

Fig. 14.—Sectional Elevation of Turkey-red Stove.

pass first through channels made of fire-brick, then through iron pipes, and finally make their exit to the chimney at B. The upper part of the stove is divided, by floors of iron-grating, into several storeys C, D, E, F, each of which is furnished with wooden framework, supporting, one above the other, two pairs of horizontal rails provided with short, upright, wooden pegs. Over these pegs one selvedge of the cloth is firmly hooked alternately from right to left, while the other is allowed to hang down; thus, when the stove is filled, each storey is closely packed with two tiers of cloth suspended in such a manner that the heated air from below can readily pass between each fold.

A yarn stove is similarly constructed, but in this case the ends of the rods holding the yarn are supported on horizontal rails free from pegs.

Another mode of hanging cloth, but one which is not so economical of space, is to have only one storey in the stove. Above, near the roof, are fixed a number of strong, smooth, wooden rails, on which the cloth is suspended in long folds, reaching down to within one or two feet of the iron grating immediately above the hot flues.

In all cases efficient ventilation is secured by means of numerous side windows, which can be readily opened and closed at will.

10*th Operation Steeping.*—Run the cloth in the open width through a machine consisting of a large vat divided into several compartments fitted with rollers above and below. The first compartments are filled with a solution of carbonate of soda at $\frac{1}{2}°$ Tw. (Sp. Gr. 1·0025), and heated to 40° C. The last is filled with water only.

The cloth is then well washed, and dried in the stove at about 65° C.

11*th to* 14*th Operation.*—These operations, consisting of mordanting, dyeing, and clearing, are precisely similar to those already described for yarn-dyeing.

It may be well to state that the number of paddings in dilute soda solution (liquoring) varies according to the quantity of oil which it is desired to fix upon the cloth. Good Turkey-red contains about 10 % of modified oil on the fibre.

"*Sulphated Oil Process*" *for Dyeing* 500 *kg.* = 9·8 *cwt. of Yarn or Cloth.*—In this process the frequent repetitions of passing the fabric through oil-emulsions or sodium carbonate and then stoving are not used. The olive oil is replaced by an alkaline solution of sulphated olive or castor oil, with which only a single impregnation is necessary, followed by a steaming or stoving process.

1st Operation: Bleaching or Boiling.—This is identical with that already given in describing the previous processes for yarn and cloth.

2nd Operation: Preparing. — The dry cotton is thoroughly impregnated by " tramping " or " padding " with a cold or tepid solution of 10-15 kg. = 22·0-33·0 lb. of neutralised sulphated-oil (50 %) per 100 litres = 22·0 gals. of water. The excess is removed, and the cotton is merely dried in the stove, or it may be heated for 1-2 hours at 75° C.

3rd Operation: Steaming.—The prepared and dried cotton is submitted to the action of steam, 2-5 lb. pressure, during 1-1½ hour.

4th Operation: Mordanting.—The cotton is worked and steeped for 2-4 hours in a tepid solution of commercial aluminium acetate (tin-red-liquor), or more economically in basic aluminium sulphate, $Al_2(SO_4)_2(OH)_2$, at 8° Tw. (Sp. Gr. 1·04).

After mordanting, the excess of aluminium solution is removed by wringing or hydro-extracting, the cotton is dried, and then either simply well washed in cold water or first worked for ½ hour, at 40°-50° C., in a chalk bath containing 20-30 g. = 308·6-462·9 grains of ground chalk per litre (1 quart). A solution of sodium phosphate may replace the chalk water. Alkaline fixing-agents like ammonia and sodium carbonate are best avoided in case any of the oil-preparation should be stripped off.

5th Operation: Dyeing.—Dye with 15-20 % of Alizarin (10 %), with the addition of 1 % of its weight of chalk or acetate of lime. The cotton is dyed in the cold for ½ hour to ensure regularity of colour, the temperature is then gradually raised to 70° C. in the course of 1 hour, and the dyeing is continued at this temperature till the bath is exhausted. The cotton is then well washed (although with highly calcareous water this is best omitted), hydro-extracted, and dried.

6th Operation: Second Preparing.—The dyed and dried cotton is again impregnated with a dilute solution of neutralised sulphated oil (namely, 50-60 g. = 771·6-925·9 grains of sulphated oil [50 %] per litre), and then dried. This second preparing may also take place after the mordanting, the oil being then fixed by means of a second mordanting with a weak solution of basic aluminium sulphate, etc.

7th Operation: Second Steaming.—The dried cotton is steamed as before for 1 hour.

8th and 9th Operations: First and Second Clearing.— These may be identical with operations 13 and 14, described in the " Emulsion process," although many chemists think that soap alone should be used here, and consider that the addition of stannous chloride is altogether unnecessary if not irrational.

The " sulphated-oil process " is comparatively so new that numerous slight modifications of the process as here given are naturally tried and adopted by various dyers, and to some of these reference will now be made.

The sulphated-oil used is invariably carefully neutralised, either with caustic soda or ammonia. As a rule, ammonia is preferred, since even the addition of an excess of ammonia would have little or no injurious effect, owing to its volatility; and further, the ammonia compound of sulphated-oil is more readily decomposed on steaming than the sodium compound, and a more complete fixing of the oil results. Either sulphated castor oil or olive oil may be used. Very good results are even obtained by the simple use of a carefully made castor oil soap, which, being excessively soluble, and giving thin solutions, is well fitted to impregnate the fibre thoroughly.

In the " preparing " process, the cotton does not attract or fix any of the oil. It simply absorbs a definite amount of the solution, and supposing sulphated olive oil to have been used, the prepared cotton contains the sodium or ammonium compounds of oxyoleïc acid and of the glycerine-sulphuric-ethers of oxyoleïc and oxystearic acids, these being its constituent elements. It is very important to know the exact percentage of sulphated-oil contained in the solution, since it is this which determines the amount of oil and alumina ultimately fixed on the cotton, and consequently the beauty, brilliancy, and fastness of the colour.

According to Liechti and Suida, the action of the first steaming process is to decompose the ammonium or sodium compounds of the ether constituent of sulphated oil into ammonium or sodium sulphate, glycerine, oxyleïc and

114 FABRIC DYEING & TEXTILE COLORING MIXTURES.

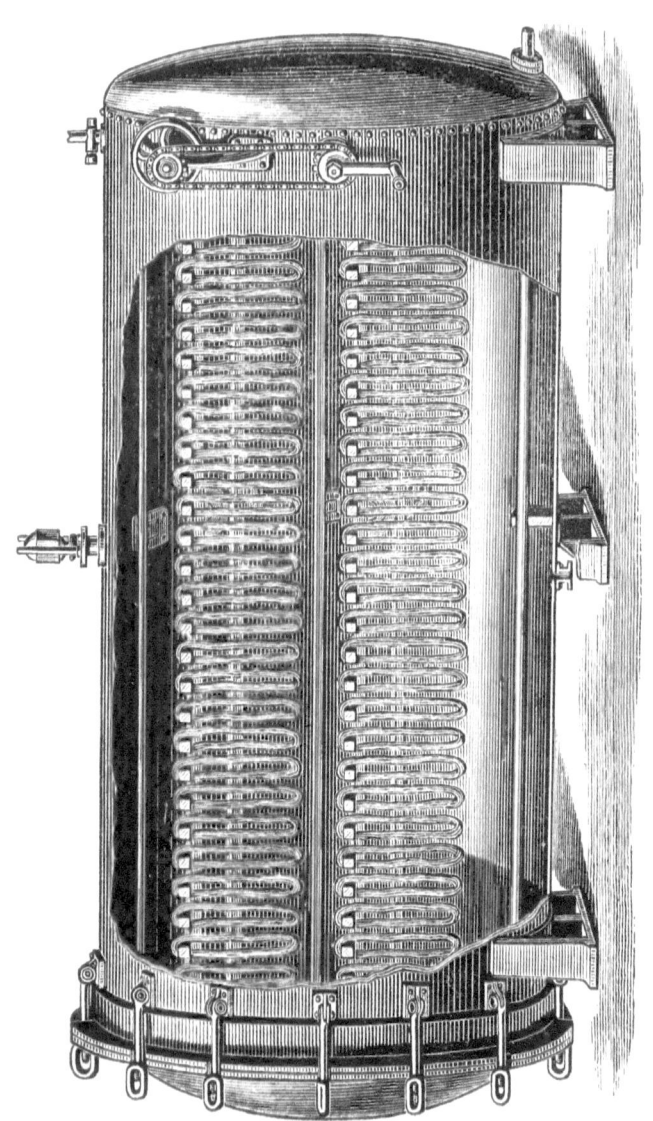

Fig. 15.—Steaming Chest for Turkey-red Yarn.

oxystearic or trioxyoleïc acid, according as olive or castor oil has been employed. The other constituent (oxyoleïc or trioxyoleïc acid) remains unchanged. At the same time the steaming causes a better penetration of the fibre by these oxidised fatty acids. Sometimes the steaming at this stage is altogether omitted. The decomposition of the compound ether referred to may also be effected by heating the dye-bath to the boiling point instead of only to 70° C., the bath becomes acid, and the brilliancy of the colour is developed suddenly.

Fig. 15 represents a steaming-chest for yarn made by Messrs. Tulpin Frères, of Rouen. The hanks of cotton are suspended on square wooden rods resting on an iron skeleton carriage or framework, and are capable of being turned during the steaming process to ensure every portion being efficiently steamed. The iron carriage is supported on wheels, so that it can be filled with yarn and then run into the chest. The steaming-chest itself consists of a wrought-iron horizontal boiler, with a movable door at one end provided with clamps. For the prevention of drops there is fixed internally and at the top a cover of sheet copper, in such a manner as to leave a space between it and the boiler-plate. The chest is provided with a steam-gauge, safety-valve, and blow-off pipe. The steam enters by a perforated pipe running along the bottom of the boiler, and which is usually covered with a perforated iron plate.

Cotton cloth may be reeled and suspended on rods in a similar way, or it may be steamed in the continuous steaming-chest of Messrs. Duncan Stewart & Co., Glasgow, represented in Figs. 16 and 17. It consists of an annular-shaped iron cylinder or chamber A B, in the upper part of which a series of brass radial rods C are caused to circulate slowly by means of the endless screw E, driven by the engine D. The cloth (in the open width) enters the annular space through a pair of squeezing rollers at F. By an ingenious arrangement the cloth, which is suspended in long, loosely-hanging folds on the radial rods, is carried round the annular space, and makes its exit by a second pair of squeezing rollers at G. The chamber is constructed of boiler-plate, so that the goods can be submitted to high-pressure steam. Another form of continuous steaming-

116 *FABRIC DYEING & TEXTILE COLORING MIXTURES.*

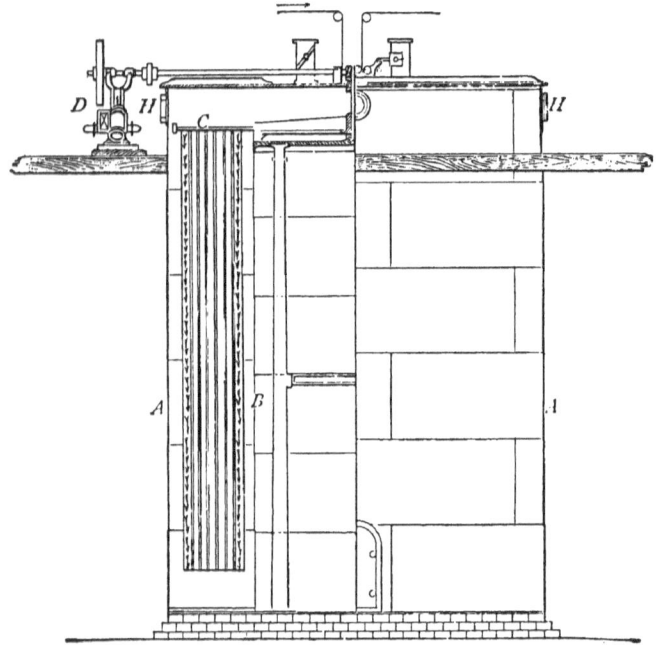

Fig. 16.—Elevation of Continuous Steaming-Chest.

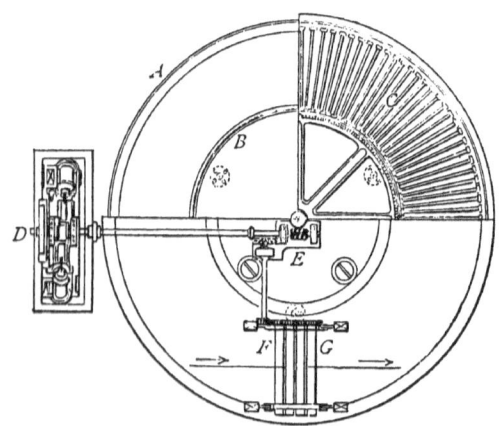

Fig. 17.—Plan of Continuous Steaming-Chest.

machine is that in which loose rods, supporting the cloth in a similar manner, are passed continuously, by means of endless chains, through a large rectangular brick chamber filled with very low-pressure steam.

If in the mordanting process the cotton was merely dried after the preparing with sulphated oil, there are produced upon the fibre the aluminium compounds, both of the ether and of the oxy- or trioxyoleïc acid; but if it was also steamed, there is then fixed on the fibre essentially the normal aluminium compound of oxy- or trioxy-oleic acid (Liechti and Suida).

A brighter colour is obtained by adding a small pro portion of stannous chloride to the aluminium solution, or stannate of soda to the oil solution.

After mordanting and washing, a slightly basic aluminium salt remains on the fibre, its basic character being generally caused by the calcareous condition of the water. Traces of lime are also present. If, previous to washing, a warm chalk bath is used, a much more basic and more calcareous aluminium compound is formed.

During the dyeing process there is probably formed the Alizarin compound of the basic oxy- or trioxy-oleate of aluminium and calcium just referred to.

If there is a deficiency of oil on the fibre, the brightest shades are always obtained by dyeing at the low temperature indicated (70° C.), but otherwise the temperature may be raised to the boiling point, although there is then a tendency of a portion of the oily mordant being softened and boiled out, especially if it is in slight excess.

With the use of pure Alizarin—*i.e.* the " blue shade of Alizarin," as it is generally called—a fiery brilliant red is not obtained; hence such as contains Isopurpurin (Anthrapurpurin)—*i.e.* the " yellow shade of Alizarin "—is generally preferred.

The second preparing and steaming operations have for their object the neutralising of the basic compound present on the fibre at this stage. This operation of steaming after dyeing has a most remarkable effect in giving brilliancy and fastness to the colour, especially if the dyeing has been conducted at a low temperature. If 100° C. was employed, then the brightening effect has taken place to a considerable extent, if not entirely, already in the dye-bath, as above mentioned. Sometimes the second preparing is omitted, and a small quantity of neutralised sulphated oil is added to the dye-bath instead.

The method of "clearing" described, in which the cotton remains stationary while the liquor circulates through it, gives very much better results than if the cotton were worked vigorously in the solution, since in this latter case much of the red-lake would be mechanically removed by friction, and the colour would look poor and weak.

The Action of Lime Salts in the Dye-bath.—One of the most interesting facts connected with the application of Alizarin is the necessity of the presence of a lime salt in the dye-bath, in order to obtain a really good, serviceable colour. The general result of researches made by Schlumberger, Rosenstiehl, and others, with a view to elucidate this point, seems to favour the idea that the Alizarin-red lake, as fixed upon textile fibres, is not simply an aluminium compound of Alizarin, but one which also contains calcium as an essential constituent.

The following results of Liechti and Suida's researches bearing on this point will explain this. Pure aluminium hydrate, whether in its precipitated form or fixed on the fibre, cannot be properly dyed with Alizarin except in the presence of lime compounds. Normal aluminium phosphate behaves similarly. Normal aluminium alizarate $[Al_2(C_{14}H_6O_4)_3]$ is a purplish-red compound soluble in water, alcohol, and ammonia. Basic aluminum alizarates —*e.g.* $[Al_2(C_{14}H_6O_4)(OH)_4]$ and $[Al_4(C_{14}H_6O_4)(OH)_{10}]$— are, on the contrary, bright red compounds, insoluble in water and alcohol, and little soluble in ammonia. Aluminium-calcium-alizarates of very varied composition, prepared by dyeing aluminium hydrate with Alizarin in the presence of calcium acetate, are mostly reddish-brown insoluble compounds. The amount of lime salt present in the dye-bath determines the quantity of Alizarin which will be taken up by the alumina, and the amount of lime taken up by the lake is determined by the quantity of Alizarin employed. The relative proportions of Alizarin, alumina, lime, and fatty acid present in the lake abstracted from Turkey-red and Alizarin-red-dyed cotton, vary considerably, according to the method of dyeing employed; as a rule, a large excess of alumina, in proportion to the lime and Alizarin, is present. The above-mentioned authors find that alumina-mordanted cotton, when dyed with Alizarin in the presence of calcium acetate, takes up one-third molecule of lime for each molecule of Alizarin, and they consider that the composition of the lake in unsoaped Alizarin-red-dyed cotton is best expressed by the formula $[Al_2Ca(C_{14}H_6O_4)_3(OH)_2]$.

Another Method of Dyeing Alizarin Red on cotton, in use among calico-printers, but which does not give quite such fast colours as those described above, is given in the following *résumé:*—

1. Mordant with commercial aluminium acetate, 5°-8° Tw. (Sp. Gr. 1·025-1·04), dry, and age for 1-2 days, by hanging in a chamber heated to 50° C., and having a somewhat moist atmosphere.

During this " ageing " process much of the acetic acid escapes, and alumina or a basic aluminium acetate is fixed on the fibre.

2. In order to fix the alumina more completely, work the cotton for a few minutes, at 60° C., in a bath of phosphate, arsenate, or silicate of soda, 5-10 g. = 77·1-154·3 grains per litre = 2·2 gals.; then wash well in water.

3. Dye with Alizarin, and dry. A small addition of acetate of lime is used if necessary. Since the cotton at this stage contains no oil, it is essential to the obtaining of a bright colour that the dyeing should take place at a temperature not exceeding 70°-75° C.

4. Prepare with a neutralised solution of sulphated oil 50-100 g. = 1·6-3·2 ozs. (50 %) per litre = 0·2 gal., and dry.

5. Steam for $\frac{1}{2}$-$\frac{3}{4}$ hour at 2 lb. pressure.

6. Clear as before.

Alizarin Pinks, Purples, etc., on Cotton.—Alizarin pinks are obtained by precisely the same methods as are adopted for reds. The aluminium mordant employed must, however, be considerably weaker—use, say, aluminium acetate at 10° Tw. (Sp. Gr. 1·05). Basic mordants are avoided, since they given uneven colours, and even normal aluminium sulphate may be used with advantage; the amount of Alizarin (20 %) may be reduced to about 1 % of the weight of cotton, and the proportion of oil preparation required is correspondingly diminished. The most pleasing pinks are those produced by using a "blue shade of Alizarin," *i.e.* one free from Isopurpurin, Flavopurpurin, or Purpurin.

Very good fast shades of purple and lilac are obtained from Alizarin, either with or without the use of oil preparation; indeed, the use of oil does not seem to add any particular brilliancy to the colour, but serves mainly to fix the mordant, and to make the colour a little faster to boiling soap solutions.

When the cotton is prepared with oil, according to either the Emulsion or Steiner's process for Turkey-red, it is mordanted, worked and steeped for a short time in a solution of ferrous sulphate at $3°-4°$ Tw. (Sp. Gr. $1·015$-$1·02$); it is then allowed to lie over-night, and is finally well washed.

The amount of iron precipitated on the fibre, and of Alizarin subsequently taken up in the dye-bath, is determined by the amount of oil previously fixed, and not merely by the concentration of the ferrous sulphate solution. For the darker shades of purple, therefore, the cotton should be well prepared with oil, while for pale shades the preparation is slight. The best and bluest shades are only obtained when the mordant is thoroughly saturated with Alizarin; excess of uncombined mordant gives the colour an unpleasant dull-reddish appearance.

The use of pyrolignite of iron gives somewhat darker, brighter, and bluer shades than the sulphate. Very deep purplish blacks are obtained, and with less oil preparation, by steeping the cotton, previous to mordanting, in an infusion of gall nuts or other tannin matter.

After mordanting, the cotton is well washed and dyed with 5-15 % of Alizarin (10 %). If the water is not sufficiently calcareous, it is very essential to add the necessary quantity of chalk or acetate of lime to the dye-bath (1-2 %). After dyeing, the cotton should be washed and soaped at a temperature of 60° C.

When not prepared with oil, the cotton is prepared with tannin, by working it in a cold infusion of tannin matter (equal to 1-2 g. = $15·4$-$30·8$ grains of tannic acid per litre = $0·2$ gal.); it is then mordanted in a solution of pyrolignite of iron, $1°-3°$ Tw (Sp. Gr. $1·005$-$1·015$), and finally washed.

One may also mordant the cotton by impregnating it with pyrolignite of iron, 1°-3° Tw., wringing out the excess, and then working it for 10 minutes, at 50° C., in a solution containing 20 cm^3 = 1·2 cubic inches of silicate of soda, 16° Tw. (Sp. Gr. 1·08) per litre = 0·2 gal., and finally washing it. Anthrapurpurin gives greyish violets, Flavopurpurin and Purpurin reddish violets, which are little esteemed.

Various shades of chocolate, claret-red, etc., are obtained with Alizarin by mordanting the oil-prepared cotton with a mixture of aluminium and iron mordants, either in the state of sulphates or acetates. Whether the cotton is prepared with oil according to the Emulsion or Steiner's process, or by the sulphated oil method, it is advisable to work it in a weak tannin bath before mordanting, especially for the darker and bluer shades, since a better proportion of iron is fixed by this means.

The different shades are produced by varying the relative proportions of aluminium and iron mordant, remembering always to vary the concentration of the tannin bath in accordance with the latter.

After mordanting, the cotton is washed, dyed with Alizarin, prepared with weak sulphated oil, steamed, and soaped, as already described.

Certain shades of claret-red may also be obtained by mordanting with a solution of chromium acetate, instead of with the mixture of iron and aluminium salts.

Although Alizarin and Anthrapurpurin have been mainly alluded to in the above, the other members of the anthracene group—namely, Flavopurpurin and Purpurin —may be applied in exactly the same way, and give rise to similar shades.

Application to Wool.—Alizarin is capable of yielding a number of pleasing shades on wool, according to the mordant used, and ought to be largely employed whenever fastness to milling and to light is required. In conjunction with other colouring matters which are similarly applied, it may yield an endless variety of shades. Its application presents little or no difficulty.

To obtain Alizarin-red on wool, mordant the wool with 6-10 % of aluminium sulphate (cake alum) and 5-8 % of cream of tartar. Introduce the wool into the cold solution, raise the temperature gradually to the boiling point in 1 hour, and continue boiling $\frac{1}{4}$-$\frac{1}{2}$ hour. Wash well, and dye in a separate bath, with 10 % of Alizarin (20 %) and 4-6 % of acetate of lime (solid). In order to ensure an even colour it is well to work the wool for $\frac{1}{2}$ hour in the cold dye-liquor, then to raise the temperature gradually, in the course of 1 hour, to the boiling point, and boil $\frac{1}{4}$ hour, or till the bath is exhausted. After dyeing, wash well, and dry at a low temperature or in the open air.

The addition of cream of tartar to the mordanting bath is absolutely essential to the production of a full rich colour. Excess of tartar tends to give intensity, but diminishes the brilliancy of the colour. Care must be taken that the aluminium sulphate used is free from iron. With a deficiency of mordant the colour lacks brilliancy and intensity, and if the deficiency is excessive only a poor, dull brick-red is obtained. With excess of mordant the colour tends to become yellower and less intense. Brighter and more orange shades are obtained by using, along with the aluminium sulphate, 1-4 % of stannous chloride, in which case a further addition of 1-4 % of cream of tartar is necessary.

The addition to the dye-bath of acetate of lime (or an equivalent amount of ground chalk) is also absolutely necessary if the water employed is not sufficiently calcareous. Without lime, the colour is poor and worthless, and the dye-bath is not exhausted; with excess, the red is darker and duller.

On comparing the colours given by the various members of the Alizarin family of colouring matters, it is found that Alizarin itself yields a very blue shade of red, or a claret-red; Anthrapurpurin, a bright red; Flavopurpurin, a somewhat duller and yellower red; Purpurin, a shade approaching that given by Alizarin, but much yellower, namely, a dull brownish-red.

To obtain Alizarin-orange on wool, mordant the wool with 5-8 % of stannous chloride (tin-crystals) and an equal weight of cream of tartar. Dye with 10 % of Alizarin (20 %) without the addition of acetate of lime. With the addition of 4-5 % of acetate of lime a bright orange-red is obtained, but without, the colour is very much yellower.

ANTHRACENE COLOURING MATTERS. 123

Excess of lime makes the orange still redder, but it is apt to be uneven. Alizarin-orange may also be dyed in a single bath. Alizarin gives a bright reddish orange; Flavopurpurin a bright yellowish-orange. The colour yielded by Anthrapurpurin holds an intermediate place; it is a bright orange; Purpurin gives a moderately bright orange-red.

Very rich claret-brown shades are obtained by mordanting the wool with 3 % of bichromate of potash and 1 % of sulphuric acid, 168° Tw. (Sp. Gr. 1·84). The addition of the sulphuric acid is beneficial, since it tends to give a somewhat yellower and fuller colour; it is not, however, absolutely essential. Dye with 10 % of Alizarin (20 %). The addition of 2-4 % of acetate of lime to the dye-bath makes the colour somewhat less yellow, or bluer, though apparently slightly less intense. Strange to say, its addition is by no means essential, as in the case of the aluminium mordant for dyeing reds. Good colours are also obtained by the single-bath method; use 1 % $K_2Cr_2O_7$ and 1 % H_2SO_4, 168° Tw. Numerous fast shades of brown, olive, purple, etc., are obtained by associating Alizarin with such colouring matters as Galleïn, Cœruleïn, and many of the dyewoods.

The shades yielded by the different members of the Alizarin group with chromium mordant are as follows Alizarin gives a dull purple colour; Anthrapurpurin, a much redder shade, namely, a claret-brown; Flavopurpurin, a yellower shade of claret-brown; Purpurin gives the most intense colour of all, namely, a deep claret-brown.

Very good shades, ranging from bluish-violet to slate, are obtained by mordanting wool with 4-8 % ferrous sulphate and 4-8 % cream of tartar, and dyeing in a separate bath with 10 % Alizarin (20 %) and 5 % carbonate of lime. With the single-bath method darker colours are obtained, but they are much browner and duller; use 6 % ferrous sulphate and 0·6 % oxalate of potash. Iron-alum employed instead of ferrous sulphate gives good results.

Copper sulphate as the mordant gives claret-browns, either by the mordanting and dyeing method or by the single-bath method.

With the use of ammoniacal sulphate of nickel and

uranium salts, as mordants, Alizarin yields nice shades of grey and slate.

Application to Silk.—Alizarin is little used in silk-dyeing. Good colours may be obtained by mordanting the silk according to the ordinary methods, and working it, after dyeing, in hot soap solution.

Nitro-Alizarin [$C_{14}H_5 \cdot NO_2 \cdot (OH)_2$]. — This colouring matter, also called Alizarin Orange, is produced by the action of nitrous acid on Alizarin. It is applied to the various fibres in the same way as Alizarin; although it yields fast colours, it finds as yet only a comparatively limited employment.

Application to Wool.—With aluminium mordant it yields very good orange colours. Mordant the wool with 6-8 % of aluminium sulphate and 7-9 % of cream of tartar. Excess of mordant renders the shade dull. The addition of acetate of lime to the dye-bath makes the colour browner.

With stannous chloride mordant the colour obtained varies very considerably, according to the amount of mordant employed. With a small amount (1 % of stannous chloride and 1·5 % of cream of tartar), a very reddish-orange is obtained; with double the amount, the colour becomes a yellowish-orange; with 4 % of stannous chloride, only a dull brown is obtained, the normal colour being evidently destroyed by the reducing action of an excess of mordant. The addition of acetate of lime to the dye-bath is not beneficial, since the yellowish-orange colour is thereby changed to brown.

With stannic chloride (equivalent to 6 % stannous chloride, $SnCl_2 \cdot 2H_2O$) an orange colour is also obtained. Excess of mordant does not destroy the colour as in the case of stannous chloride.

With copper sulphate mordant a very good brownish-red is obtained. Use 4-6 % of copper sulphate, without calcium acetate.

With ferrous sulphate as the mordant a purplish-brown is obtained. Use 6-8 % of ferrous sulphate, without calcium acetate in the dye-bath.

Bichromate of potash as the mordant yields brownish-reds. Use 3 % of potassium dichromate and 2 % of sulphuric acid, 168° Tw. (Sp. Gr. 1·84). When potassium dichromate alone is employed, the colour becomes darker with increase of mordant, even till 16 % be employed.

Alizarin Blue [$C_{17}H_9NO_4$].—This colouring matter, also called Anthracene Blue, is derived from Nitro-Alizarin by heating it with glycerine and sulphuric acid. It may be considered as the quinolin of Alizarin, and has in consequence both basic and acid properties. It is met with in commerce in two forms, namely, as a paste containing about 10 % of dry substance, and as a powder under the name of Alizarin Blue S. The former is insoluble in water, although certain commercial marks (WX, WR) possess some degree of solubility. The latter, which is, indeed, a sodium disulphite compound of Alizarin Blue ($C_{17}H_9NO_4\cdot 2NaHSO_3$), is readily soluble in water, with a brownish-red colour. Its solutions decompose, if heated to 70° C., with precipitation of the insoluble form of blue. With lime it forms an insoluble compound; hence the presence of lime salts in the dye-bath must be avoided, otherwise there will be a loss of colouring matter.

The insoluble form of Alizarin Blue may be applied in dyeing, according to the indigo-vat method, by reducing it with zinc powder and carbonate of soda, or by the ordinary method of mordanting and dyeing in separate baths. When the latter method is employed, a certain proportion of disulphite of soda may be added to the dye-bath to render it soluble, or the dyeing at 100° C. must be long continued. Avoid the use of copper dye-vessels.

With Alizarin Blue S the mordanting and dyeing method only is employed.

Application to Cotton.—Mordant the cotton with chromium according to the alkaline method. Dye in a separate bath with Alizarin Blue; raise the temperature gradually to the boiling point in the course of 1½ hours, and continue boiling for ½ hour.

Application to Wool.—The most suitable mordant to employ is bichromate of potash, in the proportion of 3-6 % of the weight of wool. The addition of sulphuric acid, 168° Tw. (Sp. Gr. 1·84), is not beneficial if used in larger amount than 1 %. Dye in a separate bath with Alizarin

Blue; raise the temperature gradually to the boiling point, and continue boiling until a bright pure shade is obtained. With insufficient boiling the colouring matter is only superficially attached to the fibre. The colour obtained is a bright indigo-blue, with purplish bloom. It is exceedingly fast to scouring, milling, light, etc., and has the advantage of not rubbing off.

When aluminium sulphate is the mordant employed a purplish-blue is obtained, which is very liable to be uneven unless great care is taken. Use 6-8 % of aluminium sulphate and 5-7 % of cream of tartar.

With stannous chloride mordant a much redder purple is obtained. Use 4 % of stannous chloride (crystals) and 2 % of cream of tartar. This mordant is not suitable for employing alone.

Ferrous sulphate, as a mordant for Alizarin Blue, is also little suitable. It gives a greenish-blue colour, possessing little brilliancy, and apt to be uneven. Mordant with 4 % ferrous sulphate and 8 % cream of tartar.

Application to Silk.—Mordant the silk with aluminium or iron in the usual manner; wash and dye in a separate bath with Alizarin Blue. Brighten the colour afterwards by boiling the silk in a soap bath.

CHAPTER VIII.

CHROME YELLOW—IRON BUFF—MANGANESE BROWN—PRUSSIAN BLUE.

THE mineral colouring matters applied in dyeing are extremely limited, and they are almost entirely confined to the vegetable fibres, the most notable exception in this respect being Prussian Blue, and this, strictly speaking, is not a mineral colouring matter.

Chrome Yellow.—Reference has already been made to the production of this colour in describing the application of bichromate of potash and of lead salts to the cotton fibre. In addition to the methods there indicated, the following, specially intended for orange, may be used:—

Prepare a bath of plumbate of lime by adding a solution of 15-25 kg. = 33·0-55·0 lb. of pyrolignite of lead to milk of lime containing 20-30 kg. = 44·0-66·1 lb. of lime, and 500 litres = 110·0 gals. of water. The mixture is well agitated, and then allowed to settle for about 2 hours.

The cotton is worked, and steeped in the more or less milky supernatant liquid for 1-2 hours, then squeezed and washed. Dye in a cold or tepid (40°-50° C.) solution containing 5 % of bichromate of potash, and ½-1 % of sulphuric acid, 168° Tw. (Sp. Gr. 1·84). Wash, and develop the orange colour by passing the cotton into clear, boiling lime-water, then wash and dry. The cotton must be removed from the lime-water bath whenever the full orange colour is developed, otherwise the colour loses brilliancy.

Iron Buff.—This colour simply consists of ferric oxide. It is produced by first impregnating the cotton with a ferrous salt solution, then passing it through an alkaline solution, to precipitate ferrous hydrate; the latter is then changed into ferric hydrate by simple exposure to the air, or, preferably, by passing the cotton into a cold dilute solution of bleaching-powder.

Instead of a ferrous salt, one may also employ a ferric salt, as ferric sulphate or nitrate. The cotton is simply impregnated with the ferric solution, then squeezed, and passed rapidly through a dilute solution of carbonate of soda, ammonia, or milk of lime. In this case, ferric hydrate is at once precipitated on the fibre, and no subsequent oxidation is necessary.

Iron Buffs are very fast to light and boiling alkaline solutions, but are sensitive to the action of acids.

Manganese Brown.—The production of this colour on cotton has been already briefly described; it is exactly analogous to the production of Iron Buff from ferrous salts. The process is simplified by adding a little sodium hypochlorite to a solution of caustic soda, passing the cotton impregnated with manganous chloride at once through this mixture. In this case, precipitation and oxidation take place simultaneously.

It is very important always to use caustic soda free from carbonate, otherwise a little manganous carbonate is precipitated on the fibre, and since this compound does not oxidise readily, the colour is apt to be irregular.

According to A. Endler, irregularity of colour may also arise from the unsuitable physical properties of the precipitate itself, when it is produced in the ordinary manner described. Endler obviates these defects by passing the cotton, after impregnation with manganous chloride, into a bath containing 25 litres = 5·5 gals. of water, 7 litres = 1·5 gals. of ammonia, and 500 g. = 1·6 ozs. of bichromate of potash. A somewhat unstable chromate of manganese is formed on the fibre, which, on decomposing, allows the chromic acid to react on the manganous hydrate and change it into some higher state of oxidation. A final passage in dilute bleaching-powder solution completes the process.

Manganese Brown is very fast to the action of light, alkalis, and acids.

Prussian Blue: Application to Cotton.—Prussian Blue was formerly very much dyed upon cotton. Since the introduction of Aniline blues it has been much less employed.

The cotton is first dyed an Iron Buff, and is then dyed in a cold solution of potassium ferrocyanide, 20 g. = 308·6 grains per litre, with the addition of 10 g. = 154·3 grains of sulphuric acid, 168° Tw. (Sp. Gr. 1·84). Wash and dry. The intensity of the blue depends upon the quantity of ferric oxide fixed upon the fibre in dyeing the buff.

Fine purplish shades of blue are obtained by working the cotton at 30° C. in nitro-sulphate of iron at 5° Tw. (Sp. Gr. 1·025), to which 2-3 % of stannous chloride has been added, and then dyeing in a cold acidified solution of potassium ferrocyanide. Wash and dry; or, if a still more purplish tone of colour is required, work for a short time in a tepid bath containing Methyl Violet or Logwood liquor.

Alkaline or boiling soap solutions readily decompose Prussian Blue, leaving brown ferric oxide on the fibre. Prolonged exposure to sunlight causes the blue to fade, but it is restored if kept for some time in the dark.

Application to Wool.—Prussian Blues (sometimes also called Royal Blues) are obtained on wool by means of red and yellow prussiate of potash, *i.e.* potassium ferri- and ferrocyanide. The former gives the best results.

The method depends upon the fact that when a mineral acid is added to solutions of either of these salts the corresponding hydro-ferri- or hydro-ferrocyanic acids are liberated, and these, under the influence of heat and by oxidation, decompose and produce insoluble Prussian Blue. If, then, wool is boiled in an acidified solution of these salts, the liberated acids are taken up by the wool, decomposition takes place gradually, and Prussian Blue is precipitated, and becomes fixed on the wool.

The wool is introduced into a cold bath containing a solution of 10 % of red prussiate of potash, and 20 % of sulphuric acid, at 168° Tw. (Sp. Gr. 1·84); the temperature is gradually raised in the course of 1 hour to 100° C., and this temperature is maintained for $\frac{1}{2}$-$\frac{3}{4}$ hour.

The colour is rendered brighter and more purplish by adding 1-2 % of stannous chloride during the last $\frac{1}{2}$-$\frac{3}{4}$ hour of the boiling.

Although sulphuric acid gives the best result, one may also use nitric or hydrochloric acid, in which case the shade of blue is modified slightly. Nitric acid, for example, makes the shade greener. It is very usual with dyers to employ a mixture of all three acids, especially when yellow prussiate of potash is employed. This mixture of acids, which is called "royal blue spirits," or merely "blue spirits," may vary slightly in composition with different dyers. A usual mixture is the following: 4 measures of sulphuric acid, 168° Tw. (Sp. Gr. 1·84); 2 measures of hydrochloric acid, 32° Tw. (Sp. Gr. 1·16); and 1-2 measures of nitric acid, 64° Tw. (Sp. Gr. 1·32).

When yellow prussiate of potash is employed, the use of nitric acid gives the best result, probably by reason of its oxidising action. For 10 % of yellow prussiate of potash use 8-12 % of nitric acid, 64° Tw.

Instead of stannous chloride in the crystalline state, the dyer generally uses it in solution, as "muriate of tin." It is often sold to the dyer as "finishing blue spirits," though under this name it generally contains a slight addition of sulphuric or oxalic acid, or both. These additions, however, are not essential.

Another method of dyeing Prussian Blue, but one now seldom employed, is the following:—

The wool is worked for 2 hours, at 30° C., in a solution of ferric sulphate, 2° Tw. (Sp. Gr. 1·01), containing 2-3 % of stannous chloride, and 2-8 % of cream of tartar. The material is then well washed, and worked for 2-3 hours, at 80°-90° C., in a bath containing 1 % of yellow prussiate of potash, and 4 % of oxalic acid, or sulphuric acid, 168° Tw. (Sp. Gr. 1·84).

In the first bath there is fixed on the wool ferric oxide, which combines with the free hydro-ferrocyanic acid contained in the second bath. The depth of blue is regulated by the strength of the ferric sulphate solution, and the amount of yellow prussiate in the second bath should correspond to the amount of ferric oxide fixed upon the wool.

Application to Silk.—Prussian Blue is now seldom dyed on silk, except as a groundwork for black.

What was formerly known as Raymond's Blue was dyed as follows:—

Work the silk in basic ferric sulphate (nitrate of iron), 5° Tw. (Sp. Gr. 1·025), for $\frac{1}{4}$ hour, wring out, and let it lie over-night. Wash well, and work for $\frac{1}{4}$ hour in a boiling soap bath containing about 10 % of soap; wash, and dye, at 40°-50° C., for $\frac{1}{4}$-$\frac{1}{2}$ hour in a fresh bath containing 9 % of yellow prussiate of potash, and 12 % of

PRUSSIAN BLUE.

hydrochloric acid, 32° Tw. (Sp. Gr. 1·16), and finally wash well.

So-called "Napoleon's Blue" is a brighter blue, produced as follows :—

Work for ½ hour in a cold bath containing 50 % of basic ferric sulphate, 50° Tw. (Sp. Gr. 1·25), 10 % of stannous chloride, and 5 % of sulphuric acid, 168° Tw. (Sp. Gr. 1·84); wring out, wash, and work for ½ hour, at 40° C., in a second bath containing 10 % of yellow prussiate of potash, 2-5 % of red prussiate of potash, and 12-15 % of sulphuric acid, 168° Tw. (Sp. Gr. 1·84). After wringing out from this second bath, the whole process is repeated. Previous to drying, the silk is softened and brightened by working it for ½ hour in a cold bath containing an imperfectly made sulphated oil. For 1 kg. of silk, use a mixture of 150 g. = 1314·6 grains of olive oil and 15 g. = 231·4 grains of sulphuric acid, 168° Tw. (Sp. Gr. 1·84).

CHAPTER IX.

METHOD OF DEVISING EXPERIMENTS IN DYEING.

Experiments with Catechu.—In the present chapter it is intended to give some idea of the manner in which the intelligent textile colourist proceeds, in order to discover the best methods of applying colouring matters to textile fibres.

As an interesting example, the fixing of Catechu on cotton may be taken :—

First of all, the solubility of the colouring matter in ordinary solvents should be determined· in water, in acetic acid, in alkalis, etc. The behaviour of the solutions towards ordinary practically useful reagents should then be ascertained.

An aqueous solution of Catechu thus tested would, for example, show the following properties :—

Gelatin gives a voluminous reddish-coloured precipitate.

Alkalis give the solution a brownish coloration.

Lime-water gives a yellowish coloration, and a precipitate.

Aluminium salts cause the solution to become lighter-coloured and yellowish.

Ferrous salts impart an olive-green coloration.

Ferric salts impart a dark-green coloration.

Copper sulphate gives an olive coloration.

Copper acetate gives a copious dark-brown precipitate.

Lead salts give a yellowish-grey precipitate.

Potassium dichromate gives a copious brown precipitate; etc. etc.

If a piece of calico is impregnated with an aqueous solution of Catechu, then dried, and at once washed, most of the colouring matter will be removed; if, however, previous to washing, it is allowed to hang for a lengthened period, or, better still, if it is steamed, it will be observed that a portion of the colouring matter

EXPERIMENTS IN DYEING. 133

will become oxidised and thus fixed on the fibre. This takes place still more largely if a weak alkaline solution of Catechu be employed, or if some oxidising agents—as certain copper salts—be added to the aqueous solution. The best results, however, are obtained by passing the steamed calico through a solution of potassium dichromate, a fact which was already indicated by the result obtained in the last-mentioned of the examinations in the test-tube. Greenish tones of colour are obtained by passing the cloth afterwards into iron solutions, or by padding the white calico in an acetic acid solution of Catechu containing ferrous sulphate, drying, and steaming. If in this last case the steamed calico be further passed into potassium dichromate solution, the colour is greatly developed in intensity, and becomes browner.

From the foregoing preliminary experiments, the method of applying Catechu to calico might be formulated as follows: Pad the cotton in an aqueous, alkaline, or acetic acid solution of Catechu, dry, age, steam, pass through a solution of potassium dichromate, and wash, if brown tones of colour are required. For obtaining greenish tones, the method to be adopted would be to pass the calico, after steaming, through iron solutions, or to add ferrous sulphate to the padding solution.

But, however interesting the observations just recorded may be, only a very small portion of the problem has thus been solved; the question still remains, what relative and absolute proportions of the various ingredients should be employed, in order to obtain the most satisfactory result? Further, what are the best conditions under which the several substances must be applied, as regards temperature of solutions, duration of steeping, steaming, exposure to air, etc.

To answer these questions, another series of experiments is necessary. Several solutions are made, containing 25, 50, 100 g. = 385·7, 771·6, 1543·2 grains of Catechu per litre of hot water, in each of which a "swatch," "fent," or "sample" of calico of suitable size is padded, then dried, and steamed. In a similar manner several solutions of potassium dichromate are made, containing, say, 5, 10, 20 g. = 77·2, 154·3, 308·6 grains of $K_2Cr_2O_7$ per litre = 0·2 gal., and portions of each padded sample are passed into each dichromate solution, for, say, 2 minutes at the ordinary temperature. Similar portions are passed through identical solutions at a medium temperature, say 50° C., and others again at the boiling point.

The experiments can be further extended by substituting alkaline chromate solutions for the potassium dichromate, or by adding varying quantities of some iron salt —as ferrous sulphate—to the Catechu solution before padding, and then passing the cloth, as before, through acid or alkaline chromate solutions. Experiments may also be instituted to determine the best mode of applying other colouring matters in conjunction with Catechu, in order to obtain various shades of colour, etc.

Experiments with Tannic Acid.—On adding a solution of tartar emetic to a solution of tannic acid (especially if ammonium chloride be added), a voluminous white precipitate is obtained, which, on experiment, is found to possess the power of attracting most of the coal-tar colouring matters of a basic character from their solutions, and combining with them to form colour-lakes, insoluble even in soap solutions. With these premises, let it be supposed that it is the object of the dyer to determine by a series of experiments the most favourable conditions for producing these colour-lakes upon the cotton fibre.

The method of procedure would be, first of all, to prepare several tannin solutions of different degrees of concentration, containing, for example, 2·5, 5, 10, 15, 20, 25 g. = 38·5, 77·1, 154·3, 231·4, 308·6, 385·7 grains of technically pure tannic acid per litre = 0·2 gal. In each of these solutions a piece of calico should be steeped for a definite period, say 6 hours, and, after removing excess of liquid by squeezing, etc., each piece should be divided into three portions. One portion of each might be passed at once—in its moist condition—into a solution of tartar emetic; a second portion might be first dried; and a third portion might be dried and steamed previously. All the "swatches" might be immersed together in the same bath, if only one is assured that there is excess of tartar emetic present. After $\frac{1}{4}$ hour's immersion they should be thoroughly washed, to remove all loosely adhering precipitate, and then passed into a solution of some basic coal-tar colouring matter—as Methylene Blue.

EXPERIMENTS IN DYEING.

The swatches are kept constantly stirred during immersion, and are thus dyed in the cold for about $\frac{1}{2}$ hour; from time to time further slight additions of colour solution are made; the bath is then gradually heated to about 60° C., and the swatches are allowed to remain in the solution until they have acquired the maximum intensity.

They are now taken out and well washed, and each swatch is divided into two parts. One half is dried, the other is worked for about $\frac{1}{4}$ hour in a soap-bath, heated to, say, 60° C., then rinsed in water and dried. Of course, each separate swatch is marked by holes cut near the edge, in order to distinguish it from the rest, and eventually they should all be pasted in a pattern book, and references to each should be written at the side.

If all the different coloured swatches are then carefully compared with each other, the following determinations may be made :—

1. The amount of tannic acid necessary to employ for obtaining a definite shade of blue.

2. The beneficial effect produced by drying or steaming of the tannin-prepared cotton is quantitatively determined, and it is at once seen how defective is the method usually employed of passing the material in its damp condition direct into the tartar emetic bath. The difference is specially noticeable in the soaped swatches.

As to the proper concentration of the tartar emetic bath to be employed, this cannot be authoritatively determined on the small scale, but, of course, the aim must always be to thoroughly fix, or render insoluble, the whole of the tannic acid present in the fibre. By a few experiments on the large scale it will be easy to find what amounts of tartar emetic must be used, in order to leave as little excess as possible in the fixing-bath. If the goods pass through the bath in a continuous manner, they are, of course, only immersed for a very short time, and the solution must then be made more concentrated, and continuously replenished from a stock solution, in order to ensure the presence of a slight excess of tartar emetic. Only by experiments made on the large scale in the dye-works is it possible to determine also what evil effects are produced by the passage of a large number of tannin-prepared pieces through the same tartar emetic bath.

The addition of such salts as ammonium chloride, etc., to the fixing-bath, for the purpose of facilitating precipitation, may also be tried.

One point still remains to be determined, namely, the proper temperature of the fixing-bath, and this may be ascertained by an additional series of experiments, similar to those instituted in the case of fixing Catechu by means of potassium dichromate.

Stannic chloride, zinc acetate, etc., also precipitate tannic acid from its solutions, and if on economical grounds it is desired to substitute these for tartar emetic, the most favourable conditions of concentration, temperature, etc., with respect to these, must be determined according to the above method. Finally, parallel experiments must be made with the best methods of employing the different fixing-agents, as previously determined, and the most convenient may then be selected.

Experiments with Colouring Matters.—Should the problem to be solved relate to the application of a coal-tar colour to wool or silk, one must determine whether the dyeing should take place in a neutral or acid bath. In the case of silk, one may try the utility of adding " boiled-off " liquor to the bath, and, having determined the best amount to use, one proceeds, in the event of an acid bath being required, to determine the proper kind and exact amount of acid to add. One may also try the use of acid salts—alum, cream of tartar, etc.—instead of free acid. Here, too, experiments must be made to see whether it is better to dye in a cold or hot bath, or how high the temperature should be raised, and in what time the highest limit of temperature should be attained, etc.

In some cases, an alkaline dye-bath may be found the best—as with Alkali Blue—on wool and silk. Since both these fibres are more or less attacked by alkalis, especially under prolonged influence and at a high temperature, etc., it becomes imperative to make experiments for the purpose of choosing the least injurious form and the minimum amount of alkali, also the proper temperature, duration of boiling, etc.

In studying the application of such colouring matters as require the aid of mordants, the necessary experimental work becomes even more difficult than in the cases already

EXPERIMENTS IN DYEING.

cited. The following considerations, for example, require to be taken into account 1, the particular kind and amount, or concentration, of mordant to employ; 2, the conditions of mordanting—duration, temperature, etc.; 3, the fixing of the mordant. As to the subsequent dyeing, the experiments partake of the character of those already mentioned.

As an illustration of the kind of experiments pursued in respect of mordants, the application of an aluminium salt to cotton and to wool will now be sketched.

Experiments in Mordanting Cotton.—Cotton is scarcely ever mordanted with aluminium sulphate, but rather with aluminium acetate, for reasons which have already been given in the chapter on Mordants. The compound $Al_2SO_4(C_2H_3O_2)_4$ will be here considered.

Several solutions containing this compound may be made, and of such concentration, for example, that they contain what is equivalent to 10, 20, 40, 60, 80, 100 g. = 154·3, 308·6, 617·2, 925·9, 1234·5, 1543·2 grains of $Al_2(SO_4)_3 \cdot 18H_2O$ per litre = 0·2 gal.

Separate pieces of calico or cotton yarn are impregnated as evenly as possible with each of these solutions, and are then exposed for about 2 days to a moist, warm atmosphere in an ageing chamber, in order to allow the acetic acid to evaporate. After such treatment there remains on the fibre a basic salt, of the composition $Al_2O_3 \cdot SO_3$, which, neglecting the water of hydration, may also be represented thus: $Al_2O_3 \cdot 3SO_3 + 2Al_2O_3$. From this it becomes evident that for full and complete precipitation on the fibre, the latter should be passed through a weak alkaline bath, in order to remove the sulphuric acid still present.

For this purpose, one may compare the use of silicate, arsenate, phosphate, and carbonate of soda or ammonia, or simply chalk suspended in water, etc. Of these salts, it is advisable to employ solutions of various degrees of concentration, but containing equivalent amounts of substance. Such several solutions are used both cold and at a temperature of 50°-60° C. The mordanted cotton is well worked therein for 2-3 minutes, until thoroughly wetted, then washed well, and properly dyed, using a slight excess of colouring matter, as Alizarin.

It is advisable to dye for some time in the cold solution, then to raise the temperature slowly, say in the course of 1 hour, to 60° C. If the swatches seem not to take up any more colouring matter—if the mordant is saturated —they are well rinsed in water, and one half of each is moderately well soaped. After drying, the various swatches are compared with each other as to colour, and with a little practice one is soon able to determine which fixing-bath gives, relatively, the best result. In using Alizarin, it is well to remember that the dye-bath must contain a certain percentage of calcium acetate.

Experiments in Mordanting Wool.—Since the method of mordanting wool differs from that employed for cotton, experiments on the application of mordants to this fibre assume another form. With regard to aluminium mordants, for example, the ordinary plan is to boil the wool with a solution of alum or aluminium sulphate, with the addition of cream of tartar, and one has to determine the relative and absolute amounts of these constituents to be employed, in order to give, for example, the best red with Alizarin.

For this purpose, six mordanting-baths are prepared, each containing, say, 1 litre = 0·2 gal. of distilled water, and such amounts of $Al_2(SO_4)_3 \cdot 18H_2O$ as are equal to 2, 3, 4, 6, 8, 10 % of the weight of wool employed. It is convenient to take 10 g. = 154·3 grains of wool for each vessel. The mordanting-baths are then simultaneously heated, so that their temperature may be raised to the boiling point, say in the course of 1 hour, and the wool is boiled for $\frac{1}{2}$ hour longer. The swatches are then well washed and dyed simultaneously in separate baths, with equal weights of Alizarin.

The dyeing is conducted in a manner similar to that given for cotton, but it is here necessary that towards the end of the operation the wool be boiled. The addition of calcium acetate to the dye-bath is also necessary. If excess of Alizarin has been used in the dye-bath, it should be removed by boiling the dyed swatches in distilled water.

After drying, the swatches are compared with each other, and the necessary amount of aluminium sulphate to employ is fixed upon, say 6-8 %. A second series of mordanting experiments should then be made, in which equal weights of wool are mordanted, say with 6 % of aluminium sulphate alone, and also with the addition of increasing amounts of cream of tartar. It is convenient to consider the amount of aluminium sulphate employed as representing 1 molecule of the salt, and to add the cream of tartar to the several baths, in the proportion of 0, 1, 2, 3, 4, 6 molecules. The method of mordanting and dyeing is conducted exactly as already described, and on comparing the colours of the dyed swatches, it is easy to determine whether or not the addition of cream of tartar to the mordanting-bath has been at all beneficial, and, if so, which amount gives the best result.

Since, in the case of wool and silk, the mordant may be applied before, after, or simultaneously with the colouring matter, other series of experiments must be carried out in order to determine which of these methods is the best.

In comparing the dyed swatches, not only is the intensity, purity, brilliancy, regularity, etc., of the colour taken into account, but also the effect produced upon the fibre, and the behaviour of the colour towards washing, soaping, scouring, milling, rubbing, light, etc.

In all cases of experimental dyeing, indeed, it is essential to conduct the experiments under conditions as similar as possible to those which are met with on the large scale, and in judging of the results, great care must be taken to avoid the possibility of referring any effect produced to more than one cause at a time.

To sum up the whole system of experimenting on the application of colouring matters by dyeing, it may be said that, in order to determine the effect of each particular ingredient used, the dyer must perform simultaneously two or more distinct experiments, in which equal weights of the same textile material are submitted to all the necessary operations under precisely the same conditions, except as regards the amount employed of the ingredient whose action is to be studied.

Whatever, indeed, be the factor the influence of which is to be determined, whether it be the duration or temperature of mordanting or of dyeing, the character or amount of the several ingredients employed, and so on, that factor alone is varied, while the others remain unchanged. In this way a systematic series of dyeing experi-

140 FABRIC DYEING & TEXTILE COLORING MIXTURES.

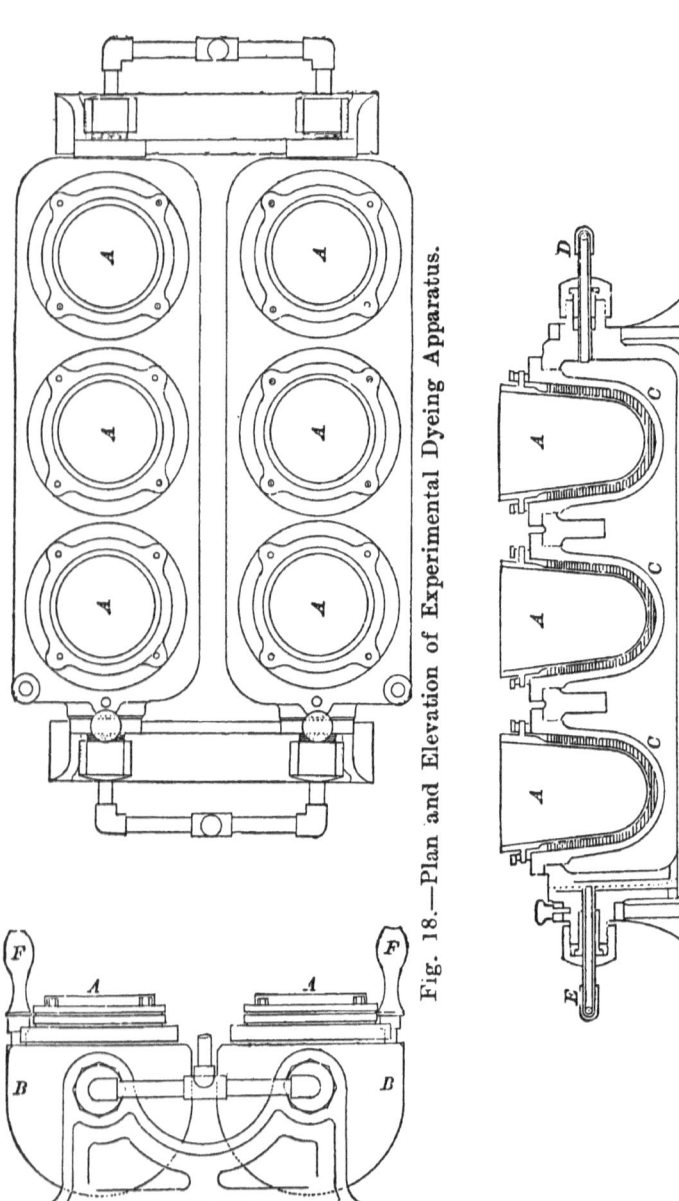

Fig. 18.—Plan and Elevation of Experimental Dyeing Apparatus.

Fig. 19.—Elevation of Fig. 18.

Fig. 20.—Section of Fig. 13.

ments is carried out; one by one the nature and value of each individual influence is carefully ascertained, until at length, by a cumulative process, the totality of conditions necessary to produce the best results is accurately determined. The actual number of experiments which it may be requisite to perform with any given colouring matter, before arriving at a full knowledge of its dyeing properties, is quite indefinite, and is more or less influenced by the character of the colouring matter, and by the general and special chemical knowledge of the experimenter.

Far too frequently, it is to be feared, the dyer neglects or declines to institute dyeing experiments on the above lines, but regards them either as of little use, or as too costly, or as taking up too much time, ignoring or forgetting that the knowledge thus obtained invariably leads to almost absolute certainty in carrying out dyeing operations on the large scale, and repays a thousandfold the time and trouble expended.

As to the apparatus required, it is comparatively simple. A water- or steam-bath, or an oil- or glycerine-bath, heated with gas or steam, provided with a perforated cover for the reception of the dye-vessels, serves for the simultaneous and equable heating of the latter. The dye-vessels themselves should be of toughened glass or well-glazed porcelain, and capable of holding about $\frac{1}{2}$ litre or even 1 litre = 0·2 gal. Metallic vessels of whatever kind, although very useful for special work, are not to be recommended for general use, especially for experimental wool-dyeing, since the acids and acid salts so frequently used in mordanting and dyeing invariably dissolve traces of the metal which, in many cases, affect the ultimate result. For the same reason, stirring-rods or other apparatus for moving the textile material during the mordanting or dyeing process, must also be of glass or porcelain. A good balance, a few glass beakers, porcelain basins, measure glasses, burettes, pipettes, and hydrometers, complete the equipment.

Figs. 18 to 20 show plan, elevation, and section of a very convenient arrangement for experimental dyeing (made by Messrs. Broadbent & Sons, Huddersfield). It consists of a couple of strong cast-iron pipes B, into which iron cups C, for holding glycerine, are screwed. The porcelain dye-vessels A rest in the glycerine cups, and are clamped down by means of a flange, protected with india-rubber or asbestos rings. The pipes B are so supported that they can be readily turned on their axles, by means of the handles F, for the purpose of emptying the dye-vessels. The axles are hollow, and serve respectively for the introduction of steam at D, and the escape of condensed water at E. One of the difficulties hitherto encountered in experimental dyeing arrangements is that of being able to heat the water in a series of porcelain dye-vessels to the boiling point simultaneously, and in a sufficiently convenient and cleanly apparatus. The one here shown, in which steam at 50-60 lb. pressure is used, will be found to answer every requirement.

Exposure of Dyed Patterns to External Influences.—Although in the above reference has been made to the method of determining the conditions of dyeing, etc., necessary to obtain the brightest, most intense, and best colours, it must not be forgotten that a dyer knows but half his business if he is simply acquainted with these conditions. He must not only learn how best to apply any given colour; he must also know the capabilities of each dyed colour how it withstands the action of light, milling, scouring, etc.—in short, all those influences, whether natural or artificial, to which the dyed fabric is likely to be submitted. Hence, all dyers should habitually and systematically expose portions of dyed patterns to the several influences just mentioned, and, as already stated, all care must be taken to avoid the possibility of referring any effect produced to more than one cause at a time. Such exposed patterns must be afterwards carefully compared with the original patterns as dyed.

Fastness of Colours.—The term " fast colour " generally implies that the colour in question resists the fading action of light, but it may also imply that it is unaffected by washing with soap and water, or by the action of acids and alkalis, scouring, milling, rubbing, bleaching, etc. In its wide sense, it means that the colour is not affected by any of those influences to which it is destined to be submitted, but its technical meaning is often restricted. To the cotton-dyer, for example, it may refer chiefly to washing with soap and water, and to light. To the

EXPERIMENTS IN DYEING. 143

woollen-dyer it may refer to milling, scouring, and light; and so on.

Many colours may be fairly fast to washing with soap and water, and yet be very fugitive towards light; or they may be fast to light, and yet very sensitive to the action of acids or alkalis.

The term "loose colour" generally implies that the colour is much impoverished, or even entirely removed, by washing with water or a solution of soap; it may, however, also mean that it is not fast to light.

The word "permanent," as applied to colour, generally denotes that it is fast to light and other natural influences.

A "fugitive colour" is generally understood to be one which is not fast to light, or which volatilises more or less under the influence of heat.

In the absence, then, of any definite meaning being attached to the above terms, it becomes imperative, in speaking of the fastness of a colour, to refer specially to the particular influences which it does or does not resist.

Influence of Light on Dyed Colours.—The chemical activity of the sun's rays is well known, and it has already been noticed that certain unstable mordant solutions seem to be decomposed and precipitated more readily under the influence of light. It is not surprising, therefore, to find that light should also have a very marked effect upon dyed colours. Under the prolonged influence of light and air almost all colours fade, and according to their relative behaviour in this respect, they are broadly divided into two classes, namely, those which are "fast to light," and those which are "not fast to light." There is, however, no definite line of demarcation between the two, and dyed colours are met with possessing all possible degrees of resistance.

Each of the coloured rays of the spectrum possesses a different fading power. White light is the most active, then follow the yellow, blue, green, orange, violet, and red rays. Direct sunlight is more energetic than diffused daylight. The light of the electric arc acts in the same sense as sunlight, but is less powerful (about one-fourth).

According to Chevreul, the presence of oxygen and moisture assists very materially in the fading action of light, so that even some fugitive colours, dyed, for example, with Safflower, Annatto, Orchil, do not fade if exposed to light in dry oxygen or *in vacuo*. Chevreul has shown, too, that the nature of the fibre has considerable influence in the matter, and that some colours are less fugitive on cotton than if fixed on wool or silk. Whether the essential action of light is one of oxidation or of reduction, or whether the action varies with each colouring matter, has not yet been determined. In the case of Prussian Blue, it is said to be one of reduction, but from the fact that air and moisture play generally an important part in the fading process, it is quite conceivable that in many cases it is one of oxidation. It is known that during the evaporation of water, ozone or hydrogen peroxide is found in small quantities, and both are powerful oxidising, as well as bleaching, agents.

Heat is found to act in some cases in the same sense as light, but in a very inferior degree.

The following notes referring to the fastness of colours are taken from experiments made with dyed wool, samples of which were exposed to the light during periods of 1, 2, 4, 8, and 12 months.

The division of the colouring matters into "fast," "medium," and "fugitive" is more or less approximate, and is merely intended to convey in a simple manner the general character of each.

Red Colours.

Fast. Alizarin (Al), Isopurpurin, Purpurin, Flavopurpurin, Nitro-alizarin (Cu), Madder.

Medium. Cochineal (Sn, Al), Biebrich Scarlet and allied colours.

Fugitive. Many of the Azo Reds and Scarlets, Magenta, Safranine, Aurin, Eosin, and allied colours; Peachwood (Al, Sn), and allied red woods; Barwood (Al, Sn), Sanderswood (Al, Sn), Ammoniacal Cochineal (Sn).

Orange and Yellow Colours.

Fast. Iron Buff, Chromate of lead yellow, Canarin, Chrysamin on cotton (Bayer), Orange G (M. L. & B.), Nitro-alizarin (Sn, Al), Alizarin (Sn), Isopurpurin (Sn), Flavopurpurin (Sn).

Medium. Croceïn Orange, Diphenylamin Orange, β Naphthol Orange, Azoflavin, Brilliant Yellow, Fast Yellow, other coal-tar yellows, Weld (Sn), Old Fustic (Sn), Quercitron Bark (Sn), Flavin (Sn), Persian Berries (Sn).

Fugitive. α Naphthol Orange, Chrysoïdine, Phosphine, Fluoresceïn, Turmeric, Annatto, Young Fustic (Sn).

Green and Olive Colours.

Fast. Cœruleïn, Naphthol Green, Persian Berries (Cu).

Medium. Weld (Cu, Fe), Old Fustic (Cu, Fe), Quercitron Bark (Cu, Fe), Flavin (Cu, Fe).

Blue Colours.

Fast. Vat Indigo, Alizarin Blue, Prussian Blue.

Medium. Logwood (Cu, Fe, Cr), Indulines.

Fugitive. Alkali Blues, Soluble Blues, Spirit Blues, Indigo Carmine, Methylene Blue (this is very much faster on cotton), Logwood (Al).

Purple Colours.

Fast. Alizarin (Fe, Cr), Isopurpurin (Fe), Galleïn (Cr, Cu).

Medium. Galleïn (Al, Fe), Cochineal (Cr, Fe), Gallocyanin.

Fugitive. Logwood (Sn), Ammoniacal Cochineal (Al), Orchil, Limawood (Cr, Fe), Methyl Violet, Hofmann's Violet, Perkin's Violet, Rosaniline Violet.

Brown Colours.

Fast. Nitro-alizarin (Cr), Isopurpurin (Cr, Cu), Flavopurpurin (Cr, Fe, Cu), Purpurin (Cr, Fe, Cu), Madder (Cr, Fe, Cu), Cochineal (Cu), Catechu.

Medium. Camwood (Cu, sadden).

Fugitive. Camwood (Cr, Cu, Al, mordant and dye), Barwood (Cr, Cu), Sanderswood (Cr, Cu), Bismarck Brown, other Azo Browns.

Some colouring matters—as Alizarin—give fast colours with all mordants; others—as Limawood and Young Fustic

—seem only capable of yielding fugitive colours; others again—as Logwood—give fast or fugitive colours, according to the mordant employed. The fugitive character of the colours obtained from Logwood by the use of tin and aluminium mordants, compared with the medium fastness of those obtained when copper, chromium, or iron mordants are employed, is rather striking.

Some colours present somewhat abnormal properties. Wool mordanted with aluminium and tin mordants, and dyed with Camwood, yields reddish-brown colours, which during exposure become at first considerably darker, and begin to fade only after 2 or 4 months. The olive-green colour yielded by Persian Berries and copper sulphate is quite remarkable in this respect, since it actually becomes darker and greener, even after an exposure of 12 months. The pure greenish-yellow obtained with Picric Acid exhibits a similar character; on exposure it rapidly becomes orange, and this begins to fade only after a lapse of about 12 months.

The mode of application also influences the fastness of the colour. Camwood and Catechu, for example, yield faster colours with copper sulphate by the saddening method than by the mordanting and dyeing method.

In studying the behaviour of the coal-tar colours towards light, one cannot fail to be struck with the manifest influence of their chemical constitution in the matter. All those colouring matters, for example, in which the atomic arrangement is like that of Magenta, are similarly fugitive to light—as Methyl Violet, Benzaldehyde Green, etc.; such similarity extends even to Aurin and Safranine. Colouring matters allied to Alizarin, on the other hand, all possess the quality of fastness to light. There are, however, cases in which an apparently slight difference in constitution gives rise to remarkable differences in fastness to light—compare, for example, Fluoresceïn and Galleïn, Indigo Carmine and Vat Indigo Blue.

The popular fallacy that coal-tar colours are fugitive, and that the colours yielded by dyewoods are fast, has already been shown to be false, and is only referred to because it still lingers in the minds of some dyers. The origin of a colouring matter has, of course, nothing whatever to do with its properties these are mainly, if not

EXPERIMENTS IN DYEING. 147

entirely, governed by its chemical composition and constitution.

Dyeing Compound Shades.—Comparatively few of the colours met with on dyed fabrics result from the employment of a single colouring matter. Therefore it becomes imperative for the dyer to know how to apply two or more together. This knowledge can readily be gained by making a series of dyeing experiments. The result obtained by mixing, as it were, the dyed colours must be observed and studied much in the same way as the artist does with the pigments upon his palette.

None of the colours here dealt with are pure, like those of the physicist; hence the product of the mixture of dyed colours is for the most part totally different from what would be produced by combining together the various colours of the spectrum.

A mixture of red and yellow produces orange, yellow and blue produce green, blue and red produce violet, green and violet produce blue. Orange and green tend to produce yellow; violet and orange tend to produce red.

The compound colours, orange, green, and violet, vary in shade according to the amount and purity of tone of each constituent single colour. If, for example, in combining yellow and blue, the yellow inclines to orange, or the blue inclines to purple, the green produced inclines to olive.

With the physicist, white is produced either by a mixture of all the colours of the spectrum, or by mixing together what are known as "complementary colours," for example, the following:—

> Purple and green.
> Red and bluish-green.
> Orange and turquoise blue.
> Yellow and ultramarine blue.
> Yellowish-green and violet.

With the dyer, however, the opposite effect tends to be produced. A judicious mixture of red, yellow, and blue tends to produce sombre colours or even black. In the same way red and green produce chocolate or brown, blue and orange produce drab, etc.

It is beyond the scope of this manual to discuss the law of the mixture of colours. For information on this point the reader is referred to Bezold's "Theory of Colour," Rood's "Modern Chromatics," and other similar works. It is, however, always imperative, in the end, to gain positive and reliable information by actually making special dyeing experiments, and even then, long experience is required before one feels thoroughly at home in producing any given compound shade.

A very necessary, or at least desirable, point to remember is that all the colouring matters employed simultaneously should be really applicable to the best advantage by the same process. A colouring matter which requires to be applied in an acid-bath ought not to be applied simultaneously with one which dyes best in a neutral bath. Basic colouring matters, although not requiring mordants, can, however, be frequently employed along with such as do, whenever the "mordanting and dyeing method" is used, since the latter are almost invariably applied in a neutral dye-bath.

If the compound shade is intended to be fast towards any influence—as light, milling, etc.—then each constituent colour yielded by the several dyestuffs, when separately employed, should be as similar in fastness to that influence as possible.

Although dyers frequently apply fast and fugitive colours together in producing compound shades, or for the sake of improving the brilliancy of any given colour, it is always more or less irrational, and ought to be avoided whenever possible.

Influence of Milling.—The process of "milling," so much used in the heavy woollen trade for Tweeds, etc., consists in saturating the woollen cloth with a strong solution of soap (frequently carbonate of soda as well) and then submitting it to a violent kneading, beating, or pressing, in the wash-stocks or "milling machine." It is an exceptionally severe treatment, and demands of the colour that it shall withstand rubbing, that it shall not be decomposed by or be soluble in weak alkalis, and that whatever colour does rub off, this shall not permanently stain contiguous fibres (bleeding). As a general rule, the best colouring matters in respect of the last point are those which require the aid of mordants, as Alizarin, Logwood, etc. Coal-tar colours, which are applied directly—as Magenta, Azo Scarlets, etc.—are very prone to dissolve off and dye the neighbouring fibres.

Some acid-colours are unsuitable because they are more or less decolorised by the action of the alkali, as Acid Magenta, Acid Green, Alkali Blue, Alkali Green, etc. In such cases the original colour can be more or less restored by a passage through dilute acid, preferably acetic acid.

CHAPTER X.

ESTIMATION OF THE VALUE OF COLOURING MATTERS.

IN order that the dyer may produce good and regular work, it is essential that the conditions under which he labours shall be maintained as regular as possible. The mordants and colouring matters he employs should be of good quality, free from injurious admixture, and of constant composition.

The certainty of obtaining such requisites can only be relied upon by exercising constant care and supervision in their selection, which should always be based on the results of analysis or practical experiment.

The purity and value of mordants can, as a rule, be determined by the ordinary methods of chemical analysis; not so, however, with colouring matters, since their analytical behaviour has been for the most part neglected.

Colorimetry.—The rapid determination of the comparative colouring power of any given colouring matter, by means of the colorimeter, has only partially and inadequately solved the problem, and it has in most cases little practical value.

The instrument consists of two calibrated tubes of equal diameter and length, into which are put the solutions of equal weights of the two colouring matters to be compared with each other. As a rule, one solution appears darker than the other, and this must be diluted with some of the solvent until both solutions are equal in intensity of colour. The colouring powers of the two dyestuffs are proportionate to the volumes of the diluted solutions.

One difficulty which presents itself is that a fair comparison can only be made when the colouring matters under examination are free from coloured impurities, and are exactly of the same tint.

Comparative Dye-Trials.—The most practical and satisfactory method of estimating the relative value of colouring matter is to make a series of comparative dyeing experiments on a small scale.

In these experiments the fibre dyed should be the same as that to which eventually the colouring matter is to be applied, and, as far as possible, exactly the same process of dyeing, and all necessary subsequent operations should be adopted as are employed on the large scale.

When it is desired to choose the best sample from a number of colouring matters which are offered for sale, equal weights of cotton-, wool-, or silk-yarn are dyed with equivalent values of the several samples, in which case that sample which gives the best results is of course the cheapest, whatever its actual price may be.

It is advisable that comparative dye-trials of this kind should be made with all the colouring matters in general use, and samples of the best and cheapest (*i.e.* those adopted) should be carefully preserved in stoppered bottles, to serve as " standards," or " types."

Aged or moist dyewoods (*e.g.* Logwood) should be previously dried, to prevent deterioration. It afterwards becomes necessary to test in a similar manner all subsequent deliveries of each colouring matter, and to compare them with their respective " standards."

One essential condition of success in making comparative dye-trials is that the several portions of yarn or cloth should be dyed simultaneously, and under exactly the same conditions as to time, temperature, etc. The most convenient dye-bath arrangement is that consisting of six or eight glazed porcelain or hardened glass dye-vessels, each of $250°$-$1000°$ cm^3 = $15·2$-$61·0$ cubic inches capacity, immersed in a common water-, oil-, or glycerine-bath, suitably heated by gas or steam, or the arrangement illustrated on p. 140.

Ground dyewoods, after weighing, may be at once put into the dye-vessels, but it is best first to prepare solutions of extracts, pastes, or soluble colouring matters, particularly if of high colouring power (as coal-tar colours). For example, $0·1$-1 g. = 2-$15·4$ grains is dissolved in 100 cm^3 = $6·1$ cubic inches of water or alcohol, and the requisite quantity of solution is introduced into the dye-vessel by means of a pipette.

Swatches of cloth are distinguished from each other by

cutting small holes near the edges; yarn is marked by means of knotted threads. After filling the dye-vessels with cold water first, the colouring matter is introduced, then the cloth or yarn, previously wetted with water. It is well, as a rule, to dye for some time in the cold, in order to ensure perfectly even dyeing; and then, with constant stirring, to raise the temperature gradually, or exactly as one would do on the large scale.

After dyeing, the swatches are washed and dried; or if necessary a portion of each may be worked for ½ hour or so in a hot soap solution, or otherwise cleansed and brightened, and then washed and dried.

The patterns are finally compared with each other under exactly similar conditions of illumination. Blue, violet, and green colours are frequently more accurately judged of by gaslight.

For those engaged in cotton-dyeing, it is convenient to employ in the above trials calico printed with various mordants, in the form of broad stripes—as with iron and aluminium mordant separately—each in two degrees of concentration, and also with a mixture of the two. Such mordanted calico was originally in general use for testing the commercial value of Madder, but it can be very conveniently used in testing all those dyestuffs which are used in practice, in conjunction with the above mordants, for the production of serviceable colours, as the various dyewoods or their extracts, Alizarin, Galleïn, etc.

For those colouring matters which can be fixed by means of tannic acid, it is convenient to use calico padded or printed in stripes with tannic acid, and afterwards fixed in a tartar emetic bath, washed and dried.

For the woollen- and silk-dyer, it is best to dye equal weights of these materials, either in form of yarn or cloth. Whenever necessary they are first mordanted, and all the ingredients which it is necessary to employ on the large scale are added to the dye-bath.

The most difficult colouring matters to estimate are those which cannot be exhausted in the dye-bath; with such, dyers often dip white blotting-paper or calico, as equally as possible, into the solutions. It is allowed to drain, and then dry, and a comparison of the colours on the stained paper is made.

VALUE OF COLOURING MATTERS. 153

In certain cases, fractional dyeing gives information as to foreign matter or impurities present in a colouring matter. A concentrated dye-bath is prepared, 0·1 m. = 4 inches of cloth is dyed in it, and after its removal a second, third, fourth, etc., are dyed in the remaining solution, until the bath is exhausted. If the colouring matter is pure, the different swatches should not vary in tone, although they may do so in intensity.

The purification of certain colouring matters by the manufacturer entails considerable loss. The cheaper qualities of Rosaniline Blue always contain reddish-violet colouring matter; hence a pure blue (6 B) gives a weaker colour than the cheaper reddish-blues. For certain purposes, the latter are indeed preferable to the former, as for the production of compound shades.

The dyer is strongly recommended not to be deterred from making exact dyeing trials because of the (comparatively little) trouble they give, for he will only in this manner properly estimate the worth of his dyestuffs, and avoid irregular or bad work.

It is probable that if the testing of dyestuffs before using them were only generally adopted, the custom would cease of mixing or diluting colouring matters with useless bodies—as dextrin, sugar, sand, etc.—which has been forced upon the colour manufacturer, because of want of judgment, on the part of many dyers, who wish only to buy cheap.

In order to recognise the individual colouring matters, one makes all possible use of the reactions dependent upon their chemical properties.

Mixtures of colouring matters are recognised sometimes by the following methods : A strip of white blotting-paper is partly immersed in the solution, when it frequently happens that the different colouring matters exhibit different degrees of capillary adhesion, and different zones of colour are perceptible on the paper; or a little of the finely-powdered colouring matter is dusted over white blotting-paper, moistened with water or other suitable solvent, or it is dusted over a glass vessel containing the solvent in a perfectly still condition. If the colouring matter is a mixture, differently-coloured spots appear on the paper, or streaks of the different colour solutions are seen in the solvent. Such an appearance is well shown by coal-tar colour mixtures, intended to give indigo-blue shades : they give violet, orange, and green spots or streaks. It is also conceivable that the different constituents of a colour mixture are not all soluble in the same solvent—a fact which may also serve to differentiate them.

Chemical methods of estimating the value of colouring matters cannot always be accepted as wholly reliable. As a rule, standard solutions of oxidising agents are employed—as bleaching-powder, potassium chlorate or dichromate in acid solution, or potassium ferrocyanide in alkaline solution—and these are added until the colour is destroyed.

These and other methods will be found in works on chemical analysis.

INDEX.

Acid, Amido-azo-sulphonic, 85-88
—, Green, applied to Silk, 52
—, —, — — Wool, 52
— Magenta, 53, 54
— — applied to Silk, 54
— — — — Wool, 53, 54
—, Picric (see Picrid Acid)
—, Rosolic, 76
—, Tannic, Experiments with, 134-136
—, Violet, 59
Alizarin, 96-124
— applied to Cotton, 97-121
— — — Wool, 121-124
— Blue, 125, 126
— — applied to Cotton, 125
— — — — Silk, 126
— — — — Wool, 125, 126
—, Experiments with, 138
— Orange applied to Wool, 122
— Pinks, 119
— Purples, 119, 120
— Red applied to Wool, 121
—, —, Dyeing Cotton with, 119
Aldehyde Green, 71
Alkali Blue, 56
— — applied to Wool, 56, 57
— — — — Silk, 58
— Green, 52, 53
— Violet, 59, 60
Aluming, Process of, 102
Aluminium Salt applied to Cotton and Wool, 137
Amaranth, 93
Amido-azo-colours, 84, 85
Amido-azo-sulphonic Acids, 85-88
Aniline Black applied to Cotton, 66-71
— — — — Silk and Wool, 71
— — Dyeing Machine, 68
— — Dyes, 66-71
— Colouring Matters, 50-71
— Colours containing Sulphur, 71
— Yellow, 84
Anisol Red, 93
Annatto, 48
Anthracene Colouring Matters, 96-126
— Green, 80-82
— Violet, 79
Azarins, 89, 90
Azo Blue, 90
— Colouring Matters, 84-95
Azoflavin, 87
Auramine, 61
— applied to Cotton, 61
— — — Silk, 61
— — — Wool, 61
Aurantia, 75
Aureosin, 77
Aurin, 76
Barberry, 49
Bark, Quercitron, 48

Barwood, 47
Basf referred to, 59-62
Bayer and Co., References to, 74, 88, 89, 90, 91, 92, 94
Bengal, Rose, 77
Berlin Aniline Co., References to, 72, 87, 91, 93, 94
Benzaldehyde Green, 50
— — applied to Cotton, 51
— — — — Silk, 51, 52
— — — — Wool, 51
Benzopurpurin, 88
Benzylrosaniline Violet, 59
Berries, Persian, 48, 49
Bichromate of Potash, Experimenting with, 37
Biebrich Scarlet, 93
Black on Boiled-off Silk, 43-46
—, Bonsor's, 40, 41
—, Chrome, 35, 36
—, Dead, 35, 36
—, Direct, 40
—, Dyeing Tussur Silk, 46
— Dyes, Aniline, 66-71 (see also Aniline Black)
—, English, 44
—, Ferrous Sulphate, 39, 40
— for Hat Plush, 43
— — — Trimmings, Mason's, 43, 44
—, Heavy, 45
—, Logwood, 31-34
—, Lyons, 44
—, Mineral, 44
— on Raw Silk, 46
— Silk, Gillet's Methods of Dyeing, 43
— Souples, Fine, 45, 46
— for Tussur Silk, 46
—, Violet, 35
— for Velvets, 44
—, Watine-Delespierre's, 40
—, Woaded, 41
Blue, Alizarin, 125, 126 (see also Alizarin Blue)
—, Alkali (see Alkali)
—, Azo, 90
— Colours, Characters of, 145
— D, New, 65
—, Dark, Dyeing Cloth, 28
—, Diphenylamine, 54
—, Ethyl, 54
—, Fluorescent, 75
—, Indigo Carmine, 10
—, Logwood, 35
—, Methyl, 54
—, Methylene (see Methylene Blue)
—, Naphthol, 82, 83
—, —, applied to Cotton, 83
—, —, — — Wool, 83
—, Napoleon's, 131
—, Night, 62
—, Prussian, 128-131 (see also Prussian Blue)

Blue, Quinoline, 72
—, Resorcin, applied to Silk, 75
—, Raymond's, 130
—, Rosaniline, 54
—, —, Purification of, 153
—, Saxony, 29
—, Soluble, 55, 56
—, —, applied to Cotton, 55, 56
—, Spirit, 54
—, Victoria, 62
— for Wool, Logwood, 41, 42
Boiled-off Silk, Black on, 43-46
Bonsor's Blacks, 40, 41
Brazilwood, 47
Broadbent's Dyeing Apparatus for Experiments, 141, 142
Brook, Reference to, 58
Brown Colours, Characters of, 145
—, Fast, 92
—, Manganese, 128
—, Orchil, 89
—, Phenyl (see Phenyl Brown)
—, Phenylene, 85 (see also Phenylene Brown)
Brüning, References to, 59, 72
Buff, Iron, 127
Calvert on Sediments in Indigo Vats, 15
Campobello Yellow, 74
Camwood, 47
—, Properties of, 146
Carmine Blue, Indigo, 10
—, Indigo, 29
Casella and Co., References to, 65, 75, 82, 84, 85, 87, 91, 93
Catechu, 49
—, Experiments with, 132-134
— Tannin Matter, 32
Chevreul on Fading Action of Light, 143, 144
Chrome Black, 35, 36
— — on Cotton, Obtaining, 33
— Yellow, 127
Chrysamin, 90
Chrysoïdine, 84, 85
— applied to Cotton, 84
— — — Silk, 85
— — — Wool, 84, 85
Chrysolin Dyes, 76
Claret Red B, 92, 93
Cloth, Distinguishing Swatches of, 151, 152
—, Dyeing, 27
— —, Sulphated Oil Process for Dyeing, 111
— —, Turkey-red (see Turkey-red)
Coal-tar Colours, Influence of Light on, 146
Coccin, 77
—, New, 93
Coccinin B, 93
Cochineal, 47, 48
Cœruleïn applied to Cotton, 80
— — — Silk, 82
— — — Wool, 81, 82
— Dyes, 80-82
Colorimetry, 150

Colouring Matters, Aniline, 50-71
— —, Anthracene, 96-126
— —, Azo, 84-95
— —, Experiments with, 136
— —, Indigo, 9-30
— —, Logwood, 31-46
— —, Oxy-azo, 88-95
— —, Mixtures of, Recognising, 153, 154
— —, Natural Red, 47-49
— —, Phenol, 72
— —, Quinoline, 72
— —, Value of, 150, 154
— —, Yellow, 47-49
Colours, Fastness of, 142, 143
Complementary Colours, 147
Compound Shades, Dyeing, 147, 148
Congo Red applied to Cotton, 87, 88
— — — — Silk and Wool, 88
Cotton, Alizarin applied to, 97-121
—, — Blue applied to, 125
—, Aluminium Salts applied to, 137
—, Aniline Black applied to, 66-71
—, Auramine applied to, 61
—, Benzaldehyde Green applied to, 51
—, Chrysoïdine applied to, 84
—, Cœruleïn applied to, 80
—, Eosins applied to, 78
—, Galleïn applied to, 79
—, Gallocyanin applied to, 83
—, Hofmann's Violet applied to, 58
—, Indigo applied to, 10-17
—, — Vats for Dyeing, 11-17
—, Indulines applied to, 63
—, Logwood applied to, 31-35
—, Methyl and Ethyl applied to, 54, 55
—, — Green applied to, 60
—, Methylene Blue applied to, 71
—, Mordanting, Experiments in, 137
—, Naphthol Blue applied to, 83
—, New Blue D applied to, 65
—, Obtaining Chrome Black on, 33
—, Oxy-azo Colours applied to, 94
—, Phenylene Brown applied to, 85
—, Prussian Blue applied to, 128, 129
—, Rosaniline Violet applied to, 58
—, Safranine applied to, 65
Croceïn, Brilliant, 93
— 3 BX, 92
— Scarlet, 94
Crystal Scarlet 6 R, 93
— Violet, 62
Cyanosin, 78
Dead Blacks, 35, 36
Dimethylaniline Orange applied to Cotton, 86
— — — — Wool, 86

INDEX. 157

Diphenylamine Blue, 54
—— Orange, 86, 87
"Direct Black," 40
Durand, References to, 77, 83
Dye-bath, Action of Lime Salts in, 118
Dyed Colours, Influence of Light on, 143
—— Patterns, External Influences on, 142
Dyeing Apparatus for Experiments, Broadbent's, 141, 142
—— Cloth, Sulphated Oil Process for, 111
—— Compound Shades, 147, 148
——, Experiments in, 132-149
—— with Indigo Vat, 26-29
——, Indigo (see Indigo Dyeing)
—— Machine for Cotton Aniline Black, 68
—— Tussur Silk Black, 46
—— Woollen Cloth, 27
—— —— Material with Indigo Vat, 26
—— —— Yarn, 27
Dyes, Aniline Black, 66-71
——, Chrysolin, 76
——, Cœruleïn, 80-82
——, Eosin, 77
——, Fluoresceïn, 76
——, Galleïn, 79
——, Indophenol, 82, 83
——, Induline, 63-65
——, Oranin, 76
——, Phthaleïns, 76
——, Rosaniline, 50-63
——, Trials for, 150-154
Dye-trials, Comparative, 150, 151
Dyewoods, Ground, 151
Emulsion Process of Turkey-red Yarn Dyeing, 97-107
Endler, Reference to, 128
English Black, 44
Eosin, 77
—— applied to Cotton, 78
—— —— —— Silk, 79
—— —— —— Wool, 79
—— B N, 77
—— Dyes, 77
Ethyl Blue, 54
—— Purple, 62
Experiments in Dyeing, 132-139
External Influences on Dyed Patterns, 142
Extract, Indigo, 9, 29
Fast Brown. 92
—— Red, 92
—— Red B, 93
—— Yellow applied to Silk, 86
—— —— —— —— Wool, 86
Fastness of Colours, 142, 143
—— —— , Notes on, 144, 145
Ferrous Sulphate Blacks, 39, 40
—— —— Vat for Indigo Dyeing, 11-15
Flavaniline, 72
Flavin, 48
Fluoresceïn Dyes, 76

Fluorescent Blue, 75
"Fugitive Colour," Explanation of, 143
Fustic, Old, 48
——, Young, 48
Galleïn applied to Cotton, 79
—— —— —— Wool, 79, 80
—— Dyes, 79
Gallocyanin applied to Cotton, 83
—— —— —— Wool, 83
Garancin supplanted by Turkey-red Dye, 97
German Vat for Indigo Dyeing, 20, 21
Gillet's Methods of Dyeing Silk, 43-45
Girard, Reference to, 77
Green, Acid, 52
——, Aldehyde, 71
——, Alkali, 52, 53
——, Anthracene, 80-82
——, Benzaldehyde (see Benzaldehyde Green)
—— Colours, Characters of, 145
——, Methyl (see Methyl Green)
——, Naphthol, 75, 76
——, ——, applied to Wool, 76
Greys, Logwood, 34
Grinding Mills for Indigo, 10
Ground Dyewoods, 151, 152
Hat Trimming, Mason's Black for, 43, 44
Hawking Machine used in Cloth Dyeing, 27, 28
Heavy Black, 45
Heliochrysin, 75
Hofmann's Violet, 58
—— —— applied to Cotton, 58
—— —— —— —— Silk, 58, 59
—— —— —— —— Wool, 58
Huguenin, References to, 77, 83
Hydrosulphite Vat for Indigo Dyeing, 17, 21-25
—— —— Liquor, Preparing, 22, 23
Indian Yellow, 87
Indigo applied to Cotton, 10-17
—— —— —— Wool, 17-30
—— Carmine, 29
—— Blue, 10
—— Colouring Matters, 9-30
—— Dyeing, Ferrous Sulphate Vat for, 11-15
—— ——, German Vat for, 20, 21
—— ——, Hydrosulphite Vat for, 17, 21-25
—— —— Machine, Mather and Platt, 14
—— ——, Potash Vat for, 20
—— ——, Soda Vat for, 20, 21
—— ——, Theory of, 9, 10
—— ——, Urine Vat for, 21
—— ——, Woad Vat for, 18-20
—— ——, Zinc Powder Vat for, 15-17
—— Extract, 9, 29
—— —— applied to Silk, 30
—— —— —— —— Wool, 29, 30
—— Grinding Mills, 10

158 FABRIC DYEING & TEXTILE COLORING MIXTURES.

Indigo Substitute, 26, 37
—— Vat, 17, 18
—— ——, Calvert on Sediment on, 15
—— —— for Cotton Dyeing, 11
—— ——, Defects in, 25, 26
—— ——, Dyeing with, 26-29
Indophenol Dyes, 82, 83
Induline Dyes, 63-65
Indulines applied to Cotton, 63
—— —— —— Silk, 64
—— —— —— Wool, 63, 64
Iron Buff, 127
——, Pyrolignite of, 32
Isopurpurin, 117
Kalle and Co., Reference to, 93
Liechti, References to, 107, 113, 117, 118
—— and Suida's Experiments on Lime-salts, 118
Light, Chevreul on Fading Action of, 143
—— on Dyed Colours, Influence of, 143
Limawood, 47
Lime Salts in Dye-bath, Action of, 118
—— ——, Liechti and Suida's Experiments on, 118
Liquor-padding Machine for Turkey-red Dyeing, Stewart's, 109, 110
Logwood, Preventing Deterioration of, 151
—— applied to Cotton, 31-35
—— —— —— Silk, 43-46
—— —— —— Wool, 35-43
—— Blacks, 31-34
—— Blues, 35
—— —— for Wool, 41, 42
—— Colouring Matters, 31-46
—— Greys, 34
—— Purples, 34
—— —— for Wool, 42
"Loose Colour" Explained, 143
Lucius, References to, 59, 72
Lutecienne, 77
Lyons Black, 44
Madder, 47
Magenta, 52
——, Acid, 53, 54
Manganese Brown, 128
Mather and Platt Indigo Dyeing Machine, 14
Meister, References to, 59, 72
Metanil Yellow, 87
Methyl Blue, 54
—— and Ethyl applied to Cotton, 54, 55
—— —— —— —— —— Silk, 55
—— —— —— —— —— Wool, 55
—— Green, 60, 61
—— —— applied to Cotton, 60
—— —— —— —— Silk, 61
—— —— —— —— Wool, 60
—— Violet, 59
Methylene Blue, 71
—— —— applied to Cotton, 71

Methylene Blue applied to Wool, 71
"Milling," Influence of, 148, 149
Mills for Grinding Indigo, 10
Mineral Black, 44
Monnet, References to, 77, 78
Mordanting Cotton, Experiments in, 137, 138
—— Wool, Experiments in, 138-142
Naphthalene Pink applied to Silk, 64
Naphthol Blue, 82, 83
—— —— applied to Cotton, 83
—— —— —— —— Wool, 83
—— Green, 75-76
—— —— applied to Wool, 76
—— Orange, 90, 91
—— Yellow applied to Silk, 74
—— —— —— —— Wool, 74
Napoleon's Blue, 131
Naphthylamine Violet, 71
Neutral Red, 65
New Blue D applied to Cotton, 65
Night Blue, 62
Nitro Compounds of Phenol, 72-75
Nitro-alizarin, 124, 125
—— applied to Wool, 124
Nitrous Acid on Phenols, Action of, 75
Nopalin, 77
Oil used in Turkey-red Dyeing, 105, 107
Oil-padding Machine for Turkey-red Dyeing, Stewart's, 108
Olive Colours, Characters of, 145
Orange Colours, Characters of, 144
——, Dimethylaniline, applied to Cotton, 86
——, ——, —— —— Wool, 86
——, Diphenylamine, 86, 87
—— G, 91
——, Naphthol, 90, 91
——, Palatine, 75
Oranin Dyes, 76
Orchil, 48
—— Brown, 89
Oxy-azo Colouring Matters, 88-95
—— Colours applied to Cotton, 94
—— —— —— Silk, 95
—— —— —— Wool, 94, 95
Palatine Orange, 75
Peachwood, 47
Perkin's Violet, 63
"Permanent" Defined, 143
Persian Berries, 48, 49
—— ——, Properties of, 146
Phenetol Red, 93
Phenol Colouring Matters, 72-
——, Nitro Compounds of, 72-75
Phenols, Action of Nitrous Acid on, 75
Phenyl Brown applied to Silk, 73
—— —— —— —— Wool, 73
Phenylene Brown, 85
—— —— applied to Cotton, 85
—— —— —— —— Silk, 85
—— —— —— —— Wool, 85
Phloxin, 77
Phosphine, 62, 63
Phthaleïns Dyes, 76

INDEX. 159

Picric Acid, 72, 73
—— —— applied to Silk, 73
—— —— —— —— Wool, 72
—— ——, Properties of, 146
Pink, Alizarin, 119
——, Naphthalene, applied to Silk, 64, 65
Poirrier, References to, 77, 89, 90, 92
Potash Bichromate, Experimenting with, 37
—— Vat for Indigo Dyeing, 20
Prussian Blue applied to Cotton, 128, 129
—— —— —— —— Silk, 130, 131
—— —— —— —— Wool, 129, 130
Purple, Alizarin, 119, 120
—— Colours, Characters of, 145
——, Ethyl, 62
——, Logwood, 34
——, ——, for Wool, 42, 43
Pyrolignite of Iron, 32
Quercitron Bark, 48
Quinoline Blue, 72
—— Colouring Matters, 72
—— Yellow, 72
Raw Silk, Black on, 46
Raymond's Blue, 130
Red, Anisol, 93
—— B, Claret, 93
—— ——, Fast, 93
—— Colouring Matters, Natural, 47
—— Colours, Characters of, 144
——, Congo, 87, 88 (see also Congo Red)
——, Fast, 92
——, Neutral, 65
——, Phenetol, 93
Resorcin Blue, 75
—— —— applied to Silk, 75
Resorcinol Yellow, 89
Rosaniline Blues, 54
—— ——, Purification of, 153
—— Dyes, 50
—— Violet, 58
—— —— applied to Cotton, 58
—— —— —— —— Wool, 58
Rose Bengal, 77
Rosolane, 63
Rosolic Acid Colours, 76
Rubeosin, 77
Safflower, 48
Safranine applied to Cotton, 65
Sanderswood, 47
Saxony Blue, 29
Scarlet, Biebrich, 93
——, Brilliant, 93
——, Crocein, 94
——, Fast, 94
——, G G, 91
——, G T, 91
—— 3 R, 92
—— 4 R, 92
—— 5 R, 93
—— 6 R, 93
—— 6 R Crystal, 93
—— R R, 92
—— S, 93

Scarlet S S, 94
——, Xylidine, 91
Silk, Acid Green applied to, 52
——, —— Magenta applied to, 54
——, Alizarin Blue applied to, 126
——, Alkali Blue applied to, 58
——, Aniline Black applied to, 71
——, Auramine applied to, 61
——, Benzaldehyde Green applied to, 51, 52
——, Boiled-off, Black on, 43-46
——, Chrysoïdine applied to, 85
——, Cœruleïn applied to, 82
——, Eosins applied to, 79
——, Fast Yellow applied to, 86
——, Gillet's Methods of Dyeing, 43-45
——, Hofmann's Violet applied to, 58, 59
——, Indigo Extract applied to, 30
——, Indulines applied to, 64
——, Logwood applied to, 43-46
——, Methyl and Ethyl applied to, 55
——, —— Green applied to, 61
——, Naphthalene Pink applied to, 64, 65
——, Naphthol Yellow applied to, 74
——, New Yellow applied to, 75
——, Oxy-azo Colours applied to, 95
——, Phenyl Brown applied to, 73
——, Phenylene Brown applied to, 85
——, Prussian Blue applied to, 130, 131
——, Raw, Black on, 46
——, Resorcin Blue applied to, 75
——, Tussur, Black Dyeing of, 46
——, Viridin applied to, 52
Simpson, Reference to, 58
Soda Vat for Indigo Dyeing, 20, 21
Soluble Blues, 55, 56
—— —— applied to Cotton, 55, 56
Souples, Fine Black, 45, 46
Spiller, Reference to, 58
Spirit Blues, 54
Steaming Chest for Turkey-red Yarn, Stewart's, 115, 116
—— —— —— Turkey-red Yarn, Tulpin's, 115
Steiner's Process for Turkey-red Cloth Dyeing, 107-111
Stewart's Liquor-padding Machine for Turkey-red Dyeing, 109
—— Oil-padding Machine for Turkey-red Dyeing, 108
—— Steaming Chest for Turkey-red Yarn, 115, 116
—— Yarn Wringing Machine for Turkey-red Dyeing, 99
Stove, Turkey-red, 110
Suida, References to, 113, 117, 118
Sulphated Oil Process for Cloth Dyeing, 110
Sulphur, Aniline Colours Containing, 71
Sumaching, Process of, 102

Swatches of Cloth, Distinguishing, 151, 152
Tannic Acid, 134-136
Tannin Matter, Catechu, 32
Trapœlin, 91
Trials for Dyes, 150-154
Tropœlin Y, 89
Tulpin's Steaming Chest, 113
Turkey-red Cloth Dyeing, Steiner' Process for, 107-111
—— Dyeing, Oil used in, 105, 107
—— ——, Stewart's Liquor-padding Machine for, 109
—— ——, —— Oil-padding Machine for, 108
—— Stove, 110
—— Yarn Dyeing, Emulsion Process of, 97-107
—— —— —, Processes for, 97
—— —— —, Stewart's Steaming Chest for, 115, 116
—— ——, Tulpin's Steaming Chest for, 115
—— ——, Weser's Tramping Machine for, 99
—— —— Wringing Machine, 99
Turmeric, 49
Tussur Silk, Dyeing, Black, 46
Urine Vat for Indigo Dyeing, 21
Vat for Dyeing Indigo, 13-30
—— —— —— Cotton Indigo, 11-17
——, Ferrous Sulphate, 11-15
——, German, 20, 21
——, Hydrosulphite, 17, 21-25
——, Indigo, 17, 18
——, ——, Defects in, 25 29
——, ——, Dyeing with, 26-29
——, Liquor, Hydrosulphite, 22-25
——, Potash, 20
——, Soda, 20, 21
——, Urine, 21
——, Woad, 18-20
——, Zinc Powder, 15-17
Vat-blue, 9, 10
Velvets, Black for, 44
Victoria Blue, 62
—— Yellow, 73, 74
Violet, Acid, 59, 60
——, Alkali, 59
——, Anthracene, 79
——, Benzylrosaniline, 59
—— Blacks, 35
——, Crystal, 62
——, Hofmann's, 58, 59
——, Methyl, 59
——, Naphthylamine, 71
——, Perkin's, 63
——, Rosaniline, 58
Viridin, 52, 53
——, Application of, 52
Watine-Delespierre's Black, 40
Weld, 48
Weser's Tramping Machine, 98
Willm B. and Girard, 77
Woad Vat for Indigo Dyeing, 18-20
Woaded Blacks, 41

Wool, Acid Green applied to, 52
——, —— Magenta applied to, 53, 54
——, —— Alizarin applied to, 121-124
——, —— Blue applied to, 125, 126
——, —— Orange on, 122
——, —— Red on, 121
——, Alkali Blue applied to, 56, 57
——, Aniline Black applied to, 71
——, Auramine applied to, 61
——, Benzaldehyde Green applied to, 551
——, Chrysoïdine applied to, 84, 85
——, Cœruleïn applied to, 81, 82
——, Eosins applied to, 79
——, Fast Yellow applied to, 86
——, Galleïn applied to, 79, 80
——, Gallocyanin applied to, 83
——, Hofmann's Violet applied to, 58
——, Indigo Extract applied to, 29
——, Indulines applied to, 63, 64
——, Logwood applied to, 35-43
——, —— Blues for, 41, 42
——, —— Purples for, 42, 43
——, Methyl and Ethyl applied to, 55
——, —— Green applied to, 60
——, Methylene Blue applied to, 71
——, Mordanting, 138-142
——, Naphthol Blue applied to, 83
——, —— Green applied to, 76
——, —— Yellow applied to, 74
——, New Yellow applied to, 74
——, Nitro-Alizarin applied to, 124
——, Oxy-azo Colours applied to, 94
——, Phenyl Brown applied to, 73
——, Phenylene Brown applied to, 85
——, Prussian Blue applied to, 129
——, Rosaniline Violet applied to, 58
—— Viridin applied to, 53
Woollen Cloth, Dyeing, 27
—— ——, Hawking Machine used in Dyeing, 27, 28
—— Materials, Dyeing, 26
—— Yarn, Dyeing, 27
Xylidine Scarlets, 91
Yarn, Dyeing, 27
—— ——, Turkey-red (see Turkey-red)
Yellow, Aniline, 84
——, Brilliant, 87
——, Campobello, 74
——, Chrome, 127
—— Colouring Matters, 47-49
—— Colours, Characters of, 144
——, Fast, 85, 86
——, Indian, 87
——, Metanil, 87
——, Naphthol, 74
——, New, 74, 75
——, Quinoline, 72
——, Resorcinol, 89
——, Victoria, 73, 74
Zinc Powder Vats, 15-17

www.ingramcontent.com/pod-product-compliance
Lightning Source LLC
Chambersburg PA
CBHW021107080526
44587CB00010B/428